Practical Radiography

Principles and Applications

by Peter Hertrich

Publicis Corporate Publishing

Bibliographic information from Die Deutsche Bibliothek
Die Deutsche Bibliothek lists this publication in the Deutsche Nationalbibliografie;
detailed bibliographic data are available on the Internet at http://dnb.ddb.de.

This book was carefully produced. Nevertheless, author and publisher do not
warrant the information contained therein to be free of errors. Neither the author
nor the publisher can assume any liability or legal responsibility for omissions
or errors. Terms reproduced in this book may be registered trademarks, the use
of which by third parties for their own purposes may violate the rights of the
owners of those trademarks.

www.publicis-erlangen.de/books

ISBN 3-89578-210-6

Editor: Siemens Aktiengesellschaft, Berlin und München
Publisher: Publicis Corporate Publishing, Erlangen
© 2005 by Publicis KommunikationsAgentur GmbH, GWA, Erlangen
Translated by Patricia A. Callow, Rebecca L. Kelly, Dr. Christopher Kronen and
Jonathan Lack. Translation management provided by OmniLingua Inc.

This publication and all parts thereof are protected by copyright. All rights reserved.
Any use of it outside the strict provisions of the copyright law without the consent of
the publisher is forbidden and will incur penalties. This applies particularly to reproduction, translation, microfilming or other processing, and to storage or processing in
electronic systems. It also applies to the use of extracts from the text.

Printed in Germany

Preface

X-ray technology has long played a vital role in diagnostic medical imaging, and since the emergence of digital imaging in 1972, radiography has become increasingly important. In the past decade, in particular, it has gained significance due to the growing popularity of interventional procedures and breakthroughs in flat detector technology. The rapid rise of radiographic studies has also resulted in a large body of technical literature.

In 1943, Siemens Reiniger published the training manual "Röntgenlehrgang" by Erwin A. Hoxter.

In 1953, the 6th edition of what by then was known as Hoxter's "Practical Radiography" was widely in use as a review book for courses on X-ray imaging technology, especially principles and applications. The book has since been regularly revised and expanded. The previous version appeared in 1991, and served as a basis for this book.

The current edition of "Practical Radiography" has been completely revised and expanded. It is intended to provide X-ray technologists, radiologists, service technicians, developers and sales engineers with a unique and comprehensive introduction to radiographic procedures and applications.

This book deals with the physical and technical principles of radiography. Theoretical discussions have been kept to a minimum, in spite of important advances in the field. For more detailed descriptions of the basic physical processes, please refer to the relevant technical literature. The emphasis here is on the relationship to practice.

Alongside descriptions of X-ray systems currently used in radiology, the book discusses the possibilities created by modern imaging and postprocessing as well as the potential of hospital networks and the ability to communicate medical patient data on a worldwide basis. The book also describes conventional image formation processes such as X-rays with film-screen systems and image intensifiers. In addition, it provides a historical overview of the advances in radiography.

I would like to thank all those who either directly or indirectly worked on this book and contributed to its success. Special gratitude is due the marketing employees at Siemens AG, Medical Solutions, who provided me with technical literature and were always available for discussion.

Special thanks also to Dr. Alexander Cavallaro of the Radiological Department of the University Hospital of Erlangen-Nuremberg. At Siemens AG, Medical Solutions, I wish to thank Dr. Heinrich Behner of the R&D department for High-power X-ray tubes and Detlef Kuritke of Sales, who put great effort into organizing and reviewing the manuscript.

Erlangen, January 2005   Peter H. Hertrich

Contents

1 Medicine and Technology 12

2 The Physical Principles 20
2.1 How X-rays Form 21
   2.1.1 Principles 21
   2.1.2 Two types of X-rays 24
   2.1.3 Other processes in the atomic envelope 27
2.2 What are X-rays? 28
   2.2.1 Electromagnetic waves 29
2.3 X-ray Spectrum 30
2.4 Efficiency of X-ray production 32

3 The Characteristics of X-rays 33
3.1 Radiographic Imaging Processes 34
   3.1.1 The absorption of X-rays in matter 35
   3.1.2 X-ray scatter in matter 35
3.2 The Attenuation Law 36
   3.2.1 Mass coverage 37
   3.2.2 Mass attenuation coefficient 37
3.3 Hard and Soft X-rays 41
3.4 Absorption Edges 42
   3.4.1 Iodine contrast 43
   3.4.2 Edge filter 44
3.5 The Distance Law 45

4 The Quality of X-rays 46
4.1 Half-value Layer Thickness 46
   4.1.1 Homogeneity 47
   4.1.2 ICRU radiation quality 47
4.2 Filters 48
   4.2.1 Filters to improve radiation quality 48
   4.2.2 Al equivalent 49
   4.2.3 Lead Equivalent 50

4.3 Tube Voltage and Ripple 51
   4.3.1 Tube Voltage 51
   4.3.2 Ripple 51
4.4 Measuring X-Rays 53
   4.4.1 Ionization 53
   4.4.2 Measuring Procedure 53

5 Dose 55
5.1 Radiation Exposure 55
   5.1.1 Natural radiation 56
   5.1.2 Medical exposure to radiation 56
5.2 Physical Dose Quantities 57
   5.2.1 Effective dose E 57
   5.2.2 Energy dose D 58
   5.2.3 Ion dose J 58
   5.2.4 Kerma K 59
5.3 Dose Quantities in Radiation Protection 59
   5.3.1 Equivalent dose H 59
   5.3.2 Local dose 60
   5.3.3 Personal dose 60
   5.3.4 Body dose 60
5.4 Dose Rate 61
   5.4.1 Ion dose rate 61
   5.4.2 Energy dose rate 62
   5.4.3 Equivalent dose rate 62
5.5 Dose Area Product (DAP) 62
5.6 SI Units 63
5.7 Applied Radiation Protection 63
   5.7.1 Structural radiation protection, radiation protection plan 65
   5.7.2 Technical radiation protection 67
   5.7.3 Radioprotective clothing 68
   5.7.4 Radiation protection for the patient 68
   5.7.5 Radiation dose monitoring 69
   5.7.6 Organizational radiation protection 71

6 X-ray Systems for Diagnostics and Intervention 73
6.1 Imaging Procedures without X-rays 74
   6.1.1 Sonography 74
   6.1.2 Magnetic resonance imaging 74
   6.1.3 Impedance scanning 75
   6.1.4 Nuclear medicine diagnostics 76
   6.1.5 Optical imaging processes 76

6.2 Imaging Procedures Using X-rays 76
  6.2.1 Native image diagnosis 77
  6.2.2 Examinations using contrast media 77
  6.2.3 Interventional Radiology 78
6.3 Technology of X-ray Systems 79
  6.3.1 X-ray systems for the skeleton and chest 80
  6.3.2 Trauma and emergency X-ray systems 82
  6.3.3 Mobile X-ray systems for bedside images 83
  6.3.4 X-ray systems for mammography 84
  6.3.5 Dental X-ray systems 86
  6.3.6 X-ray systems for internal medicine 86
  6.3.7 X-ray systems for urology 90
  6.3.8 X-ray systems for special angiography 93
  6.3.9 X-ray systems for surgery 97
  6.3.10 X-ray systems for computed tomography 99

7 X-ray System Components 104
7.1 Basic System 105
7.2 X-ray Generator 105
  7.2.1 Multipulse generators 106
7.3 X-ray Tube 108
  7.3.1 X-ray tube 110
  7.3.2 Cathode 114
  7.3.3 Anode 116
  7.3.4 Focal spot, focus 118
  7.3.5 Rotor and stator 120
  7.3.6 The rotating-anode starter 120
  7.3.7 Other processes in the X-ray tube 121
7.4 Primary Collimator 123
7.5 Spotfilm Device and Bucky Diaphragm 125
7.6 Stands 127
7.7 Accessories 129
  7.7.1 Patient safety 129
  7.7.2 Patient comfort 130
  7.7.3 Image quality and radiation hygiene 130
  7.7.4 Synchronous technique 131
  7.7.5 Hygiene and sterile work area 131
  7.7.6 Pediatrics 131

8 Image Receptor Systems 132
8.1 Image Modes 132
  8.1.1 Analog and digital 132
  8.1.2 Direct, indirect imaging 133

Contents

8.2 Film-screen Systems 133
   8.2.1 The X-ray film cassette 134
   8.2.2 Direct images with film changer and magazine technology ... 136
   8.2.3 X-ray film 137
   8.2.4 X-ray film processing 145
   8.2.5 Intensifying screens 146
   8.2.6 Conventional film-free procedures 152
8.3 Imaging Plate Systems 153
8.4 Image Intensifier Television Systems 158
   8.4.1 Image intensifiers 160
   8.4.2 Television pickup tube 165
   8.4.3 Image intensifier indirect exposure technique 168
8.5 CCD Systems 169
8.6 Flat Detector Systems 171
8.7 Summary of image receptor systems 177
8.8 Image viewing systems 179
   8.8.1 Film viewing 179
   8.8.2 Monitors, displays 180

9 X-ray Imaging Technology 183
9.1 Radiographic Parameters 185
   9.1.1 Generator data 185
   9.1.2 The mAs product 190
   9.1.3 Automatic exposure controls 191
   9.1.4 Error display and data logs 199
   9.1.5 Tube data 199
   9.1.6 Exposure tables 205
   9.1.7 Icons for medical technology systems 215
9.2 Digital Exposure Technique 218
   9.2.1 Digitization, digital technique 218
   9.2.2 Digital radiography operating types 225
   9.2.3 Image processing in digital radiography 234
9.3 Image Quality 244
   9.3.1 Image quality requirements 245
   9.3.2 Image quality factors (characterizing parameters) 245
   9.3.3 Measures for improving image quality 252
   9.3.4 Contrast agent 262
9.4 Projections and Imaging Planes 264
   9.4.1 Projections 264
   9.4.2 Imaging planes 270
   9.4.3 Radiological positioning method 274

10 Patient Data Management 275
10.1 Saving and Archiving Patient Data 275
10.2 Patient File 276
10.3 Medical Information Science 277
10.4 Networks 278
10.5 Local Network Structures 278
   10.5.1 Topologies 279
10.6 PACS 282
10.7 Communication and Storage 283
   10.7.1 Image acquisition station 284
   10.7.2 Image analysis and processing workstations 284
   10.7.3 RAID 285
   10.7.4 Long-term archiving (jukebox) 285
   10.7.5 Spoolers 286
   10.7.6 Routers 286
   10.7.7 Film digitizers 286
10.8 Metropolitan and Wide-area Communication 286
   10.8.1 Wide-area networks 287
10.9 Hospital and Radiology Information Systems (HIS/RIS) 287
   10.9.1 HIS 287
   10.9.2 RIS 288
10.10 Standards 288
   10.10.1 ACR/NEMA 2.0 288
   10.10.2 DICOM 3.0 289
   10.10.3 Health Level 7 (HL7) 290
   10.10.4 EDIFACT 291
10.11 GIF, JPEG, TIFF Image Format Standards 291
   10.11.1 GIF, Graphics Interchange Format 291
   10.11.2 JPEG, Joint Photographic Expert Group 291
   10.11.3 TIFF, Tagged Image File Format 291
10.12 Integrating the Healthcare Enterprise (IHE) 292

11 Appendix 293
11.1 Glossary of Terms 293
11.2 Abbreviations 308
11.3 References 310
11.4 Photo credits 311

Index 312

1 Medicine and Technology

Modern medical engineering offers a wide range of technical and physical devices for diagnosis, intervention, therapy and patient monitoring, as well as the exchange of information worldwide. From the stethoscope to highly-effective analog and digital measurement and evaluation tools, a great number of these devices are imaging systems. Their diversity, with or without X-rays, is a result of individual medical needs, and sometimes economic factors. Imaging systems have been developed specifically for typical investigational methods and are never mutually exclusive, with the exception of special, differentiated diagnoses.

New methods and procedures are not an end unto themselves. They will continually develop and improve to the benefit of individuals and society.

The discovery of X-rays profoundly influenced medical diagnostics and therapy. Even today, X-rays are used in approximately 70% of examinations involving diagnostic imaging.

The birth of radiology

When the physicist Wilhelm Conrad Röntgen (Fig. 1.1a) discovered X-rays late in evening of November 8, 1895, no one could anticipate the explosive development in medical imaging for diagnostics and therapy.

The first Nobel Prize in physics was awarded in 1910 to Wilhelm Conrad Röntgen, who was born on March 27, 1845 in Lennep, now Remscheid (home of the German X-Ray Museum), and died on February 10, 1923 in Munich. Where German is spoken, the rays that he discovered are termed roentgen rays. They go by the name of X-rays internationally since W.C. Röntgen named them mystery or X-rays when he discovered them.

W.C. Röntgen held his first and only speech to the Würzburg Physics Society on January 23, 1896 by taking an X-ray of the hand of the well-known anatomist, Privy Councilor Kölliker.

The ability to look into the body of a living human being without applying the scalpel was an absolute sensation for physicians. They were the first to recognize the immense benefit that the "new rays" could bring to their discipline and people suffering from illness.

It was the birth of radiology. The discovery of these rays by physicist W.C. Röntgen led to the creation of a new medical discipline. The 20th century became the age of X-rays, setting in motion unprecedented medical and technical development.

In May 1896, the first professional journals appeared that dealt with X-ray experiments and their results. To cite an example, the doctor Heinrich Albers-Schönberg published the journal "Progress in the Field of X-Rays" (volume one appeared in 1897/98). In 1905, he founded the German X-ray Society and in 1919, he was appointed to Germany's first chair for radiology.

The early days of radiographic technology

On December 22, 1895, W.C. Röntgen took the first image of his wife's hand (Fig. 1.1b). It is entitled "Hand with Rings" and was originally in the possession of Prof. Ludwig Zehnder, a colleague and fried of Röntgen. The first medically indicated images were taken in January 1896. The exposure times were up to one hour.

In the same year, newly developed X-ray generators were manufactured to research the human skeleton. An image of a head was created in July 1896 with an exposure time of eleven minutes using an RG&S X-ray device with a spark inductor, mercury interrupter and current reverser (see Fig. 1.1c and d). By means of comparison, today's exposure times lie within the millisecond range, and the doses have been reduced to approximately 2% of what was first required, primarily due to the introduction of digital imaging.

The university machinist Erwin Moritz Reiniger (Fig. 1.1e) opened a workshop for manufacturing physical and electromedical devices on May 24, 1877 in downtown Erlangen, Germany.

In 1893, Erwin Moritz Reiniger, Max Gebbert and Karl Schall created the company "Vereinigte Physikalisch-Mechanische Werkstätten Reiniger, Gebbert und Schall oHG." More than 110 years ago, they laid the cornerstone for what became the medical engineering division of Siemens AG, Siemens Medical Solutions, established in 1969.

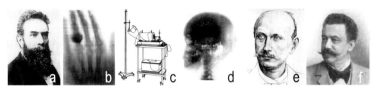

Figure 1.1 W.C. Röntgen (a); X-ray: "Hand with Rings" (b); RG&S X-ray generator (c); head X-ray, dated 1896 (d); E.M. Reiniger (e); C.H.F. Müller (f)

1 Medicine and Technology

In 1865, the glass blower Carl Heinrich Florenz Müller founded the company C.H.F. Müller. In 1899, he obtained the first patent for an X-ray tube with a water-cooled anode. In 1927, C.H.F. Müller was acquired by Philips and became its medical engineering department. Philips was founded in 1891 and manufactured carbon filament lamps in Eindhoven, Holland.

A short history of radiology

1895 Discovery of X-rays by Wilhelm Conrad Röntgen.

1896 In January, initial manufacture of X-ray generators. First X-ray tubes with a regulated vacuum (Reiniger, Gebbert & Schall).

1897 First proposal for a rotating (cathode) tube.

Radiotherapy: the therapeutic effect of ionizing radiation was used at the beginning of this century for treating different types of cancer.

First experiments involving contrast agent at Harvard, USA.

First issue of the journal, "Advances in the Field of X-Rays."

1898 Levy-Dorn: the now historical X-ray laboratory is opened in Berlin.

First use of double-cast plates with two intensifying screens.

The anticathode is angled 45° from the tube axis.

1899 Rectifying electrolytic interrupter (by Arthur Wehnelt).

1901 First X-ray exhibition involving the following companies: Reiniger, Gebbert & Schall, Siemens & Halske, C.H.F. Müller, AEG, Hirschmann and others.

Orthodiagraph to determine and record the size and shape of the heart and locate foreign bodies.

1902 Introduction of dosimetry by Holzknecht in Vienna.

Compression diaphragm (by Albers-Schönberg) for investigating objects deeply embedded in soft tissue.

1903 Recognition of the effect of X-rays on gametes by Albers-Schönberg.

Lead glass plates and protective gloves are used near the fluorescent screen to protect the investigator from X-rays.

1904 First X-ray generator with a high-voltage transformer and rotating high-voltage rectifier in X-ray technology.

This allows the negative half-waves in alternating current to be used to generate rays.

1905 Foundation of the German X-Ray Society.

Prof. Rieder of Munich reports on the use of contrast media in gastric diagnosis; the start of gastrointestinal radiography.

1906 First opaque radiograph of the kidney duct system.

X-ray tubes with a tantalum anode: The high melting point of tantalum (2,996°C) allows the X-ray tubes to withstand a higher load compared to platinum and iridium, the metals previously used as anode material.

1907 Tiltable, steel tube X-ray system for upright, angled and horizontal patient examinations.

1908 First light localizer.

A cinematograph for taking cinematic X-rays of moving body parts or other processes in the body.

1909 Tubes with tungsten anodes. Tungsten has remained the anode material of choice to this day due to its high melting point of 3,380°C and high atomic number (74), which allows it to produce superior rays (the patent was granted in 1904).

1910 Marie Curie publishes the theory of radioactivity. The contrast agent barium sulfate is introduced for gastric diagnosis.

Whole-body imaging.

1910 first electrical hearing aids.

1911 Use of the ECG with a moving-coil galvanometer.

1912 First honeycomb diaphragm by Gustav Bucky; this remains indispensable even today as a scatter radiation grid for reducing scatter radiation.

1913 Fist X-rays of mammary carcinoma in preparation.

Prof. von Laue proves that X-rays are electromagnetic waves.

1915 First three-phase generator.

First construction of a rotating-anode tube.

1921 X-ray dosimeter to determine the dose in röntgen per second (r/s).

1923 First device (by H.H. Berg) for targeted serial X-ray images.

1924 First radiological imaging of the gall bladder, bile ducts and blood vessels.

1928 Introduction of rotating anode tubes.

1 Medicine and Technology

 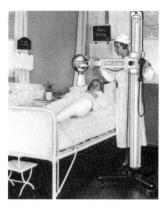

Figure 1.2 Left: Final product examination in the shipping hall in 1900. At this time, a moveable X-ray system termed an "X-ray cabinet" was already standard equipment in the examination room. Right: In 1933, an imaging system known as the "X-ray sphere" was the most sold and well known self-contained X-ray diagnostic generator. 40,000 units were manufactured over 40 years.

1929   Self-catheterization of the heart by Werner Forssmann (Nobel Prize 1956).

1933   First 400 kV X-ray tube.

"X-ray sphere" (see Fig. 1.2, right).

1935   Introduction of linear tomography.

1937   Introduction of two-button generator operation (kV and mAs) in radiological imaging.

1938   Introduction of the primary diaphragm (collimator) with a light localizer and light-beam diaphragm.

Starting in 1938, serial examinations were introduced in Germany for the early detection of tuberculosis with specially developed chest stands.

1939   Manufacture of X-ray therapy generators (Fig. 1.2, left).

1945   Visualization of coronary arteries.

1946   Introduction of automated X-ray exposure timer.

1948   First X-ray automatic exposure timer with automatic shut-off (Iontomat).

1950   Use of nuclear medicine.

1951   X-ray cinematography by R. Janker with overtable X-ray tubes.

1953	Cut-film changer and roll film changer with up to 8 images per second for angiography.
1954	Use of high-kV radiography.
1955	Introduction of X-ray image intensifier systems.
	Mid-1950s: US echo-sounding method for cardiac diagnosis by Eder and Hertz.
1957	Start of angiography on the high-pressure side of the heart using catheters through the aorta, bronchial tube and chest.
1958	Cassette-controlled automatic formatting.
	First cardiac pacemaker with a fixed beat.
1962	42 MeV radiotherapeutic machine.
	First medical use of laser beams.
1964	Percutaneous transluminal angioplasty (PTA) by Charles T. Dotter.
1966	Start of real-time ultrasound.
1967	Christiaan Barnard, South African surgeon, performed the first successful transplantation of a human heart on December 3, 1967 in Cape Town.
1970	Widespread introduction of mammography.
	First developments with CCDs.
1971	First clinical use of computed tomography of the skull by G.N. Hounsfield and J. Ambrose.
1972	Introduction of computer tomography (CT).
1975	Real-time digital subtraction angiography (DSA).
	Introduction of CCD sensors in the mid-1970s.
1976	First computer tomograph for whole-body tomography.
	Clinical use of positron emission tomography (PET).
1977	First radiofrequency generator in a mobile imaging unit.
1978	Development of digital radiography for fluoroscopic systems
1979	Use of radiofrequency generators. First 50 kW RF generator.
1980	Introduction of magnetic resonance (MR), first head images.
	First CCD area sensors with low resolution.
	In the 1980s, individual genes were isolated for the first time and their information was decoded.

1982 First 80 kW RF generator with interference-free data transmission using fiber-optic cables.

1983 First superconductive MR system with magnetic self-shielding.

1984 3D image processing.

1985 Widespread use of radiography-supported interventional techniques. Balloon catheter (Charles T. Dotter and Andreas Grüntzig).

1986 Initial testing of CCD sensors in medical diagnostics.

1988 Introduction of digital radiography (DFR).

1989 Convincing clinical results with spiral CT.

First surgical image intensifier with CCD technology. Use of CCDs in professional cameras and special high-resolution sensors.

1991 Publication of the DICOM 3.0 communication standard for the worldwide exchange of medical images.

1992 Introduction of radiology networks (PACS).

Beginning of the 1990s: selenium detectors for digital detector radiography of the chest.

1995 Clinical tests of flat detectors (20 cm × 20 cm) in cardiac systems.

1996 Clinical testing of large flat detectors (43 cm × 43 cm) in radiological systems for angiography and cardiac angiography.

1999 First routine system installation using flat detectors in X-ray rooms.

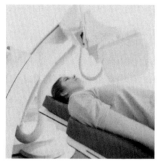

Figure 1.3 2004: Highly-modern process-oriented production line for radiography systems (left) and C-arm angiography system with a flat detector as the imaging system (right)

2001 The first transatlantic operation was successfully performed on September 7, 2001: gall bladder removal. The remote-controlled robot "ZEUS" in the Louis Pasteur University of Strasburg was operated from a control console at the Mount Sinai Medical Center in New York.

A delay of just 130 ms was attained over 7,000 kilometers using high-speed fiberglass cable.

2002 First flat detectors (Fig. 1.3) for dynamic scenes in angiography and cardiac angiography.

2003 Heart catheters guided by magnetic navigation.

2004 Rotation time of 0.37 seconds in CT.

First investigations of the small intestine using an endoscope.

2 The Physical Principles

To better understand the functional processes and technology behind modern X-ray systems, it is important to have a general knowledge of the physical laws and processes underlying X-ray generation and application. Such knowledge also helps us better appreciate the value of future developments.

The same principle since 1896

X-rays are the main data transmitters in the field of radiological diagnostic imaging. They are used to create an image of the inside of the human body.

This section discusses the general physics behind X-rays and how they interact with matter and respond to changes in physical parameters.

For detailed descriptions of the basic physical processes, refer to the relevant technical literature (see appendix).

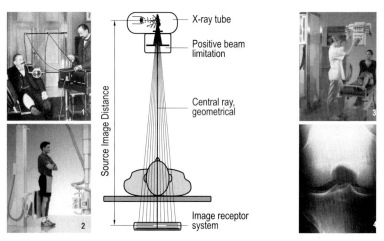

Figure 2.1 Knee images from 1900 through today (1-3), still using the same principle and result image (4).

2.1 How X-rays Form

X-rays form in a vacuum when high-energy fast electrons strike solid matter or are decelerated by some other means.

2.1.1 Principles

Figure 2.2 illustrates the principles of X-ray formation. The source of the electrons is a coiled tungsten wire (1) heated by an electrical current (I_H = heat current) to approx. 2,000°C to 2,600°C. At this high temperature, many electrons are released or emitted. The tungsten wire is also the cathode (4) of the tube, which is called the filament cathode or the negative electrode of the X-ray tube.

Figure 2.2 shows a schematic diagram of the electrons (2) emitted by the heated tungsten wire (1). They are accelerated towards the anode by the voltage applied between the cathode (4) and anode (5) (U_R = tube voltage, also potential or potential gradient) (e.g., at U_R = 100 kV to approx. 165,000 km/s). The electrons are accelerated inside a glass tube in a virtual vacuum (10^{-6} to 10^{-7} mbar), where there are no collisions to slow them down.

The X-rays originate at the point where the electrons emitted by the cathode penetrate the atoms of the anode material. This point is known as the focal spot or focus.

The area of the anode where the electrons strike and penetrate is predetermined (see section 7.3 and Figure 7.12). This point is located on the rotating anode, a rotating disk (6) used for most applications. This rotating anode is coated with a rhenium tungsten alloy and/or molybdenum or graphite (see section 7.3).

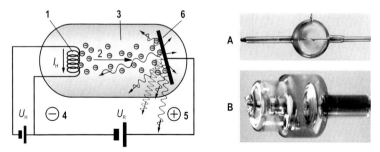

Figure 2.2 The principle of X-ray formation in a vacuum tube. Tube with stationary anode from 1915 (A) and with rotating anode from 1980 (B).

2 The Physical Principles

Understanding basic physics

The density of matter is the ratio of the mass of individual atoms to the number of atoms in the volume. When the electrons emitted by the cathode penetrate the anode, they interact with the anode atoms. The output from this interaction includes X-rays.

- The density ρ (rho) is the ratio of the mass m of a substance (matter) in grams (g) to its volume V (cm^3): $\rho = m/V$ (g/cm^3).
- Mass is what causes inertia (resistance) and all bodies to have weight (in gravity).

The Atom

The Bohr atomic model (see appendix) can be used to generally describe how X-rays originate.

Fig. 2.3 illustrates the structure of an atom. Nucleons, the positively charged protons (1a) and electrically neutral neutrons (1b), make up the nucleus of the atom (1). The nucleus is approx. 10^{-12} to 10^{-13} cm in size.

The number of protons in neutral, non-ionized atoms is equal to the number of negatively charged electrons (3) that orbit the nucleus at high speed. Electrons orbit the nucleus at various solid angles along complex pathways (2), also known as shells or atomic envelopes. The entire atom, including its electrons, is approx. 10^{-8} cm in size.

The atom as a whole has a specific mass. The energy holding all the particles of an atom together is known as binding energy E_b, which the atom itself provides. It represents only a small percentage of the atom's intrinsic energy.

The full mass of the atom, and therefore the binding energy of all the atomic particles, increases with the atomic number Z (see appendix). The binding energy of the electrons in the shells (E_1, E_2, etc.) decreases as the distance

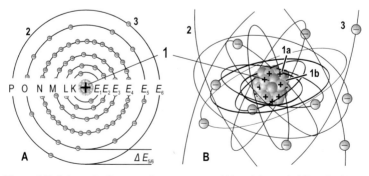

Figure 2.3 Schematic diagram of tungsten atom (A) and the probability distribution of the electrons in the atomic envelope (B)

Number of shells	K	L	M	N	O	P	Q
Number of principal quantum n	1	2	3	4	5	n.a.	n.a.
max. number of electrons per shell	2	8	18	32	50	n.a.	n.a.
Number of electrons per shell at tungsten	2	8	18	32	12	2	n.a.
Bonding energy of the electrons to the atomic nucleus (keV)	69.51	11.00	2.80	0.59	0.07	0.02	n.a.

Figure 2.4 Tungsten atom data (binding energy in keV)

from the nucleus increases. Conversely, the energy potential ΔE between shells increases the further they are from the nucleus (see the tungsten example in Fig. 2.4).

According to the Bohr atomic model, electrons orbit the nucleus in a maximum of 7 shells. Proceeding outward from the nucleus, each shell is identified by the amount of electron binding energy it contains, called the primary quantum number $n = 1, 2, 3$, etc., or K, L, M, N, O, P and Q. The electrons are identified by their shell, i.e., K electrons, L electrons, etc. (Figures 2.3 and 2.4).

The primary quantum number (Fig. 2.4) is used to determine the maximum number of electrons possible in each shell.

$$Z_{max} = 2 \cdot n^2.$$

The total number of electrons in a tungsten atom is 74 and corresponds to the number of protons in the nucleus. Tungsten is located in position 74 of the periodic table of elements. Rhenium (Re) has one additional electron in the O shell, placing it in position 75.

Electrons in the outermost occupied shell are called valence electrons. In its basic state, each atom has a unique number of occupied shells, which increases with the atomic number (in theory, an infinite number of shells are possible).

The nucleus of the atom

The total number of nucleons (protons and neutrons) in the nucleus is called the mass number. The mass of the nucleus of an atom is determined experimentally using spectroscopic methods and is expressed in atomic mass units.

Atoms with different numbers of protons and the same number of neutrons are called isotopes. The atoms all belong to the same chemical element, even through their (relative) atomic mass is different.

A nuclide is an isotope with the same number of protons and a specifically defined number of neutrons.

2 The Physical Principles

In the field of nuclear medicine diagnostics, radionuclides (isotopes with radioactive properties) are used to study bodily fluids or metabolic products to measure biochemical and physiological processes in the body (in vivo methods).

Atomic nuclei with unequal numbers of neutrons and/or protons have a gyroscopic intrinsic angular momentum (nuclear spin) that is linked to a proportional magnetic moment. The property of the various types of nuclei and their corresponding nuclear spin resonance in the human body, especially in oxygen atoms, are used in magnetic resonance imaging.

2.1.2 Two types of X-rays

The kinetic energy of electrons is converted into X-ray radiation (quantum energy, photon energy) through deceleration of the electrons emitted by the cathode in the electrical field of the anode atoms and through electron collision processes in their atomic envelopes.

The penetration depth of the electrons emitted by the heating coil and accelerated in the potential gradient is up to 30 µm. As with both types of radiation (i.e., bremsstrahlung and characteristic radiation), it depends on the anode material and the tube voltage U_R (the tube voltage potential between the cathode and anode).

- A negatively charged carrier with the mass "m" (in this example, the negatively charged cathode electron), that passes through an electrical field from "–" (cathode) to "+" (anode) (the potential gradient of the tube voltage U_R), gains kinetic energy E_{kin} (in general, the reverse is also true).
- In addition to the SI unit J (Joule, see appendix), the electron volt is the unit often used to measure energy, especially in atomic and nuclear physics; $E = 1 \text{ eV} = 1.602 \cdot 10^{-19}$ J (the elementary charge of the electron).
- An emitted cathode electron accelerated by 1 V (potential gradient) gains a kinetic energy E_{kin} of 1 eV (electron volt).
- At a tube voltage of $U_R = 100$ kV, the electrons have a kinetic energy of 100 keV (1000 eV = 1 keV = 10^3 eV, 1 MeV = 10^6 eV, etc.) when they penetrate the anode.
- As this kinetic energy interacts with the anode atoms, approx. 99% is converted to heat (measured in Ws = watt seconds). It has no value in radiography and has to be dissipated in applied X-ray diagnostics.
- Approximately only 1% of the kinetic energy of the electrons can be used to generate X-ray quanta and ultimately for radiography.
- At typical tube voltages U_R of at least 70 kV up to 150 kV, the X-ray radiation is made up of 80-90% continuous bremsstrahlung (electron deceleration) and approx. 10-20% characteristic radiation (electron collision processes).

The conversion of the kinetic energy by the emitted cathode electrons into radiation is a complex process. For this reason, we will use a simplified model to describe the formation of X-rays used in radiological imaging.

X-ray "bremsstrahlung"

Because their charge is negative, electrons from the cathode are in most cases accelerated as they enter the positively charged field of the nuclei of the anode atoms and then deflected. The acceleration and deflection generate X-ray quanta (Fig. 2.5). In this process, the electrons give up some or (very seldom) nearly all their kinetic energy ($E = h \cdot v$, see appendix) to these quanta.

The electrons lose energy to the quanta and decelerate. They enter into the electrical field of another atom (Fig. 2.5, right), are re-accelerated and again deflected. That means that the electrons emitted from the cathode usually interact the same way in sequence with many atoms of the anode material. This causes them to decelerate incrementally. The X-ray quanta generated are known as X-ray bremsstrahlung (the German word means "braking radiation"). Each deceleration produces a quantum with a different energy ($h \cdot v_{1-n}$) (see section 2.3).

- After the negatively charged electrons (2) emitted by the cathode pass through the tube voltage U_R, they are attracted to the positively charged nuclei of the anode atoms (1a, 1b), accelerated, and deflected. They lose a specific amount of their kinetic energy ($h \cdot v_{1-5}$, see appendix), depending on their distance from the nucleus. This kinetic energy is converted into X-ray quanta (3), creating heterogeneous radiation, i.e., X-rays with various wavelengths (3) independent of the atom's properties.

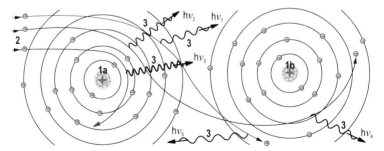

Figure 2.5 Principle of formation of X-ray bremsstrahlung through deceleration in the electrical field of the anode atoms (interaction with the nucleus of the atom)

Characteristic X-rays

Less frequently, rather than generating X-ray radiation as just described, the cathode electron releases some, or in extreme cases nearly all, of its kinetic energy to an electron orbiting the anode atom (Fig. 2.6). Either the electron is completely ejected from the atom or is bumped to the next highest shell, and therefore to a higher energy level. The resulting emission is known as characteristic or intrinsic radiation.

Figure 2.6 illustrates how characteristic X-rays form: An electron from an outer orbit moves into a free space in a shell closer to the nucleus (Fig. 2.6, 6a-c), causing an emission. This process releases the difference in the binding energies ΔE of the two shells in the form of electromagnetic radiation (quantum), that is, photon radiation characteristic of that atom: K_α, K_β and Fig. 2.7: M_α, L_α, K_α, K_β. It is depicted as a line spectrum (see section 2.3 for more information on line spectra).

The released energy E is proportional to the frequency v of this radiation. At high ΔE values in the region of the inner electron shells, it is emitted as characteristic radiation with short wavelengths.

The specific potential differences ΔE and the associated wavelength of the radiation are characteristic for each atom. This property is useful in materials identification.

- The following formula applies: $E = h \cdot v = (h \cdot c)/\lambda$; whereby h is Planck's constant, v the frequency of the quantum radiation, c the speed of light, and λ the wavelength of the quantum radiation (see appendix).

Using the binding energies of tungsten as an example (see Fig. 2.4, "The Atom"), the following energies can be calculated for the X-ray quanta generated.

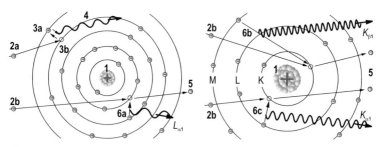

Figure 2.6 Principle of generating characteristic X-rays (interaction with shell electrons around the nucleus of the anode atoms)

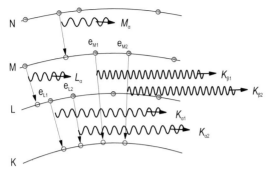

Figure 2.7 Schematic diagram of characteristic radiation forming in a tungsten anode through the transition of electrons from shells N to M, M to L, L and M to K. Characteristic photon radiation forms as well as the K_α and K_β X-rays used in radiology.

- Transition of an electron from L to K:
  K_α = 69.5 keV − 11 keV = 58.5 keV
- Transition of an electron from M to K:
  K_β = 69.5 keV − 2.8 keV = 66.7 keV

These are calculated energy values. In reality, quanta energies generated deviate slightly from the above.

- For K_α: $K_{\alpha 1}$ = 59.3 keV and $K_{\alpha 2}$ = 57.9 keV
- For K_β: $K_{\beta 1}$ = 67.2 keV and $K_{\beta 2}$ = 69.0 keV

On closer examination, the individual quanta energies (also called lines, see section 2.3) are split again according to their specific orbital states, i.e., the statistically average locations of the electrons before and after emission; Fig. 2.7: e_{L1}, e_{L2}, e_{M1}, e_{M1}. See also Fig. 2.3 for orbital states.

In the voltage range for general X-ray diagnostics higher than 70 kV, the amount of the line spectrum for characteristic radiation is approx. 10 to 20% of the total radiation. This portion has no special application in general X-ray imaging. Only in mammography does the less high-energy characteristic radiation, supported by molybdenum anodes, play a significant role (Z_{Mo} = 42; K_α radiation = 17.45 keV, K_β radiation = 19.6 keV). The imaging voltages U_R in X-ray mammography are U_R = approx. 25 to 32 kV.

2.1.3 Other processes in the atomic envelope

As previously described, X-rays are formed when the electrons emitted by the cathode penetrate the anode material. When the cathode electrons interact with the anode atoms, however, other effects also occur.

Auger effect

When an electron from an outer shell fills a vacancy in the K shell, the energy released can be passed to an adjacent shell electron, which is then ejected from the atom. This electron is known as the Auger electron. This phenomenon is an "intraatomic" photoeffect (ionization of the atomic envelope).

Photon emissions

Electrons can be excited to various states of energy in the atomic envelope. When electrons "jump" to a lower state, radiation of varying energy is released. Transitioning in the outer region of the electron envelope produces ultraviolet (UV) radiation. At lower transition energies, photon radiation in the range of visible light or infrared radiation is produced (Fig 2.6-4).

Ionization

If a non-decelerated cathode electron (Fig. 2.6-2b) collides with a shell electron of an anode atom, the latter electron may be released from its shell. For this to happen, however, the shell electron must receive the energy equal to its binding energy to the nucleus. The released electron moves freely in the material. This process is known as ionization (Fig. 2.6-5).

2.2 What are X-rays?

X-rays are electromagnetic waves just like radio waves, heat or light waves, with a constant propagation speed of $c = 2.99792458 \cdot 10^8$ m/s = speed of light (or approx. 1,080,000,000 km/h).

- Like all electromagnetic waves, X-rays are differentiated by their wavelength and the number of oscillations per second (= frequency).
- X-rays cannot be deflected with optical, magnetic or electrical fields.

As frequency increases and wavelengths become shorter, the energy of the X-ray increases. The energy of the quanta and the frequency of the waves have the following relationship: $c = \lambda \cdot v =$ constant.

c = speed of light
λ = wavelength
v = oscillations per second (frequency)

2.2 What are X-rays?

2.2.1 Electromagnetic waves

A wave is a periodic, self-propagating oscillation that is characterized by means of its wavelength (Fig. 2.8).

The wavelength λ is defined as the distance s that a full oscillation (a period T) takes. Frequency indicates the time interval T in which the oscillation takes place. Frequency is expressed in hertz (Hz, see appendix) as s^{-1}.

Figure 2.9 illustrates the range of electromagnetic waves that are relevant in medical imaging.

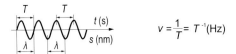

Figure 2.8 Elapsed time of a continuous sine wave

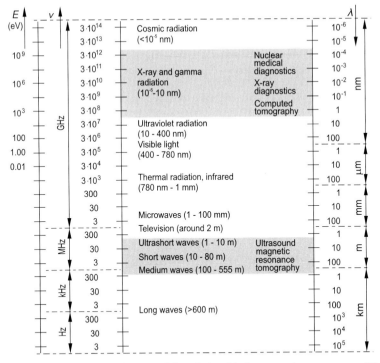

Figure 2.9 Schematic diagram of the relationship between electron energy (E), wavelengths (λ) and frequencies (v) of electromagnetic waves

Radiation, waves, quanta, photons

Lower-frequency electromagnetic radiation is usually referred to as waves, e.g., radio waves, television waves, or microwaves. For the extremely high-frequency X-rays and gamma radiation, certain phenomenon can only be described in terms of specific elementary particles, called quanta or photons.

- X-rays might simply be viewed as a large stream of quanta of varying energies moving forward in waves.
- The intensity I_ψ, or rather the energy flux density of X-rays, is the radiation field restricted to a specific energy (sum of all X-ray quanta, photons) that travels in one second through 1 cm^2 perpendicular to the propagation direction of the radiation. The intensity depends on the tube voltage U_R and the anode material.

2.3 X-ray Spectrum

The X-ray spectrum (Fig. 2.10) illustrates the intensity distribution of the wavelengths relevant for diagnostic X-rays or the frequency distribution of the quanta. It is a bremsstrahlung spectrum with a line spectrum of characteristic radiation superimposed.

In the line spectrum shown here, only individual (discrete) wavelengths or quanta energies (lines) occur. These lines, also called spectral lines, correspond to specific quanta energies (wavelengths). More precisely, they repre-

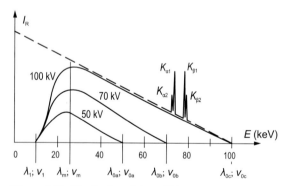

Figure 2.10 Distribution of three different continuous X-ray bremsstrahlung spectra with superimposed line spectra of characteristic radiation (K) at tube voltages (typical for diagnostic radiology) U_R of 50 kV, 70 kV and 100 kV; other typical tube voltages include 65 kV, 80 kV or 150 kV (except for mammography and endoral dental images).

sent a mixture of somewhat different energy values, so they are not lines in a strictly mathematical sense, but show a specific range (see Fig. 2.7).

Fig. 2.10: At a tube voltage of U_R = 100 kV (70 kV, 50 kV etc.), the electrons emitted by the cathode strike the anode with an energy of 100 keV (70 keV, 50 keV, etc.). Therefore, the highest-energy X-ray quanta that can form are 100-keV quanta (70-keV, 50-keV quanta, etc.). They represent the upper limiting energy of the bremsstrahlung spectrum with the maximum limiting frequency.

The maximum possible limiting energy has the highest possible limiting frequency v_{0c} (v_{0b}, v_{0a}, etc.), and the shortest possible wavelength λ_{0c} (λ_{0b}, λ_{0a}, etc.). To reach the upper limiting energy, the cathode electron must pass through the vacuum distance (potential gradient U_R) without colliding into other particles.

Then, the following applies:

$E = e \cdot U_R$; with the elementary charge of the electron e = $1.602 \cdot 10^{-19}$ As.

X-ray quanta that have the maximum possible energy are the exception. Usually, the electrons are incrementally decelerated as they interact with the anode atoms. Every deceleration produces an X-ray quantum with the energy equivalent to the specific deceleration event or the remaining released energy. Every second, billions of quanta with varying energies are created in this way. In their entirety, they form the continuous bremsstrahlung spectrum.

The dashed line (Fig. 2.10) illustrates the full bremsstrahlung spectrum created in the vacuum tube. Due to the inherent filtration in the tube, however, the lower limiting energy of the X-rays emitted is approx. 10 to 15 keV (at λ_1 and v_1). E = quanta energy; I_ψ = intensity (energy flux of the X-ray quanta as a relative measurement). It results from the absorption of low-energy X-ray quanta (approx. 10-keV to 15-keV quanta) within the X-ray tube housing and the glass envelope, if the range of the lower limiting energy is determined using the lower limiting frequency v_1, and the lower, longest wavelength λ_1.

For application in radiology, bremsstrahlung has its greatest intensity I_ψ (energy flux density of the X-ray quanta) at a mid-range quanta energy, in the range of the wavelength λ_m or frequency v_m. It is considerably lower than the upper limiting energy of the continuous bremsstrahlung spectrum.

In tungsten anodes, characteristic radiation first forms when the cathode electrons interact with the tungsten atoms at an energy > 70 keV. Energies below 70 keV are not sufficient to bump an anode electron out of the K shell in tungsten.

Depending on the tube voltage U_R and the anode material, more or less characteristic radiation is created. As previously described, the X-ray radiation emitted from the focus is usually a mixture of bremsstrahlung and characteristic radiation.

2.4 Efficiency of X-ray production

The efficiency η or yield is defined as the ratio of the X-ray radiation energy (that is ultimately available for imaging) to the full electron energy created when passing through the potential difference, the tube voltage U_R.

The efficiency η is primarily determined by the type of deceleration fields and the depth to which the cathode electrons penetrate the anode material, that is, by the atomic number Z of the anode material and the tube voltage U_R, expressed in volts (V).

$$\eta = k \cdot U_R \cdot Z$$

whereby k is a constant of magnitude (1 to 1.5)·10^{-9}. It was experimentally set to k = 1.1·10^{-9}.

For the anode material most frequently used, tungsten (Z = 74), the efficiency η is approx. 1% at a tube voltage of U_R = 125 kV. However, due to the fact that only a very small part of the radiation produced in the X-ray tube leaves the tube as a primary beam cone, the effective efficiency η is further reduced by a factor of 10.

3 The Characteristics of X-rays

X-rays yield different effects when passing through matter. The most important effects, depending on the condition of the particular matter, are described briefly below:

Luminescence effect

X-rays impinging on a fluorescent layer excite the layer causing it to emit light, e.g., in the case of intensifying screens, finger ring dosimeters, DLR (digital luminescence radiography), image intensifier input fields, and FD (flat detectors, see chapter 8).

Photographic effect

Photographic layers (emulsions) are blackened by light beams as well as by X-rays. For a comparison of the photographic effect to conventional photography, see section 8.1.

Semiconductor effect

X-rays change the conductivity and charge of semiconductors, e.g., in the case of semiconductor radiation detectors for dose rate measurement, flat detectors, etc. (see chapters 4 and 8).

Ionization effect

X-rays cause ionization in biological tissues of gases, for example. This results in changes in the electrical conductivity of the penetrated materials, e.g., in the case of equipment for measuring dose rate (see chapter 4).

Biological effect

X-rays can cause changes in living tissue. In the medical field, the indication and benefit of radiation determine the type and its application. In diagnostics, radiation protection is responsible for limiting the damaging effect radiation can have on human beings to the extent that the health disadvantages are acceptable for both the individual and society. The international radiation protection commission ICRP determines the principles of radiation protection (see chapters 4 and 5).

Given higher quantum energies, as used in radiotherapy, for example, additional interactions occur such as pair production and nuclear reaction.

Pair production

A particularly high-energy X-ray quantum with $E = h \cdot v > 1.02$ MeV (see appendix) can change in the electrical field of an atomic nucleus (internal field) of the irradiated matter into one positive and one negative electron. The quantum disappears, i.e., true absorption occurs. This case is not relevant for diagnostic X-ray radiation.

Nuclear reaction

A proton is released from the atomic nucleus during the nuclear reaction. This case is also not relevant for diagnostic X-ray radiation, since a photon energy of $E = h \cdot v$ between approx. 6 and 20 MeV is required.

3.1 Radiographic Imaging Processes

When irradiating matter (Fig. 3.1), the quanta (photons) of the X-ray radiation interact with the atoms of the matter (the human body in this instance or sections of the X-ray systems, e.g., the patient table, etc.). These are processes during which the energy or the direction of the X-ray quanta changes.

When passing through matter, the X-ray radiation generated primarily for radiographic imaging is attenuated to a greater or lesser extent depending on the matter density, thickness, etc.

Two types of attenuation processes are significant for the diagnostic X-ray radiation generated in the case of a tube voltage U_R of approx. 25 to 150 kV:

- The absorption and
- The scatter of X-ray quanta.

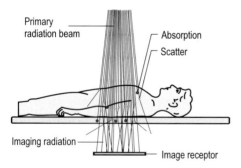

Figure 3.1 Schematic diagram of X-ray radiation penetrating matter

3.1 Radiographic Imaging Processes

3.1.1 The absorption of X-rays in matter

An X-ray quantum is absorbed (Fig.3.2) when it impinges on an electron of the atomic envelope (1) of the matter to be penetrated. For this to happen, the energy of the quantum must be greater than the bonding energy of the electron impinged on.

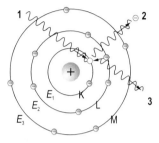

Figure 3.2 Schematic diagram of the photoeffect being generated by absorption (photoabsorption)

The electron leaves its atom and spreads throughout the matter as a free photoelectron (2). This process is known as photoabsorption (ionization), but it is generally referred to as "absorption." Of course, the absorbed X-ray quantum is no longer capable of contributing to the radiographic imaging process.

If the free space becomes occupied by an electron of an external shell, characteristic radiation can also occur during absorption of an X-ray quantum (3). This is called "fluorescent X-ray radiation."

3.1.2 X-ray scatter in matter

An X-ray quantum (photon) may interact with an electron of the atomic envelope even without or with only minimal emission of its energy.

Figure 3.3 shows the following: an X-ray quantum (1a) releases an electron from the atomic envelope of an external shell and then leaves the atom again

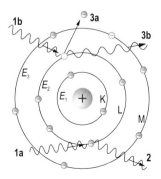

Figure 3.3 Schematic diagram of the creation of coherent (classic) and incoherent (Compton effect) scatter

at E (keV)	Proportion of quanta interaction in %			
	Absorption	Scatter	Summary	without
60	69.9	30.0	99.9	0.1
70	59.8	39.8	99.6	0.4
90	38.0	59.0	97.0	3.0

Figure 3.4 Orders of magnitude in %, how the interaction processes of absorption and scatter act in practice

in a change of direction (2). If it retains its entire energy in this process (i.e., no change in wavelength), it is a case of coherent or Rayleigh scatter (also referred to as classic scatter). Coherent scatter is predominant in the long-wave (low-energy) X-ray spectrum.

If the X-ray quantum loses part of its energy to the released electron, this effect is referred to incoherent scatter or the Compton effect. In the Compton effect (see appendix), an electron (3a) is pushed by an X-ray quantum (1b) from its shell while receiving part of the quantum energy. The scattered X-ray quantum continues on its path with the residual energy (3b). The Compton effect is important in the short-wave (high-energy) X-ray spectrum.

Fig. 3.4 shows absorption and scatter in a 10 cm layer of water which corresponds to 10 cm of tissue. Quantum energy E corresponds to the tube voltage U_R times the elementary charge of an electron.

3.2 The Attenuation Law

The attenuation law is one of the most important laws of X-ray physics. Radiological diagnostics is based entirely on the different attenuation coefficients, which depend on the quality of the X-ray radiation among other things (see chapter 4).

The attenuation law applies for intensity I, the energy flux density of the "attenuated" X-ray quanta, in this instance, after penetration of a matter:

$I = I_0 \cdot e^{-\mu d}$

I_0 = Intensity (energy flux density of the X-ray quanta) prior to penetrating the matter
μ = Linear attenuation coefficient
d = Thickness of the matter to be penetrated
e = Basis of the natural logarithm

Linear attenuation coefficient μ is a quantitative measure of the sum of all attenuating interactions. Specific coefficients are defined for the different processes of all interactions. The coefficients relevant for radiological diagnostics are absorption coefficient τ and scatter coefficient σ. σ is the sum of the scatter coefficients for coherent (classic) scatter σ_k and incoherent scatter (Compton) σ_c. Pair production coefficient χ is not relevant for imaging radiology.

Intensity I decreases exponentially according to linear attenuation coefficient μ and the above-mentioned equation.

3.2.1 Mass coverage

Instead of linear layer thickness d of the absorber material, mass coverage $d \cdot \rho$ is provided (also surface coverage; ζ). It is an SI unit and indicates the density ρ (g/cm^3) of the absorber material within which a certain thickness d (cm) is irradiated. Therefore, the mass coverage (surface coverage) is a surface-related quantity, i.e., the irradiated mass per surface unit: $d \cdot \rho$ (g/cm^2).

3.2.2 Mass attenuation coefficient

The mass attenuation coefficients are more significant than the linear attenuation coefficients (μ, τ, σ).

Since all interaction coefficients are proportional to density of the irradiated matter, linear attenuation coefficient μ relates to density ρ of the matter attenuating the X-rays. μ/ρ is the mass attenuation coefficient. Similar to linear attenuation coefficient μ, total mass attenuation coefficient μ/ρ is comprised of the coefficients of the individual interaction processes. The results are mass attenuation coefficients τ/ρ (mass absorption coefficient) and σ/ρ (mass scatter coefficient). The sum of both coefficients is mass attenuation coefficient μ/ρ.

Attenuation coefficient μ is the quotient of $(1/n_q) \cdot (\Delta n_q/\Delta d)$, where n_q is the number of quanta impinging vertically on the absorber layer of thickness Δd. Δn_q is the number of quanta interacting with the matter within thickness Δd. Since n_q is dimensionless, the dimension $\mu = 1/\text{cm}$ results.

This yields the following dimension for the mass attenuation coefficient:

- $\mu/\rho = (1/\text{cm})/(\text{g/cm}^3) = (\text{cm}^2/\text{g})$

Figure 3.5 shows the combination of mass attenuation coefficient μ/ρ (cm^2/g) for air, water (corresponds to tissue), and lead as a function of the energy of the X-ray quanta.

at E (keV)	Mass attenuation coefficients							
	Air (Z_{eff} = 7.64)			Water (Z_{eff} = 7.42)			Lead (Z = 82)		
	τ/ρ	δ/ρ	μ/ρ	τ/ρ	δ/ρ	μ/ρ	τ/ρ	δ/ρ	μ/ρ
30	0.15	0.17	0.32	0.14	0.19	0.33	24	1.35	25.35
50	0.04	0.16	0.20	0.04	0.17	0.21	5	0.65	5.65
100	0.02	0.13	0.15	0.03	0.14	0.17	5	0.26	5.26

Figure 3.5 The mass attenuation coefficient (μ/ρ) is the sum of the absorption coefficient (τ/ρ) and scatter coefficient (σ/ρ)

In the case of the quanta energy values listed, the scatter coefficient is greater than (>) the absorption coefficient for air and water; the opposite is true for lead. This means that

- Absorption < scatter for elements with low atomic numbers Z.
- Absorption > scatter for elements with higher atomic numbers Z.

Quanta energy E (keV) is similar to tube voltage U_R.

As with the creation of X-rays (X-ray quanta), complex physical processes also occur during interaction of the X-ray quanta with the irradiated matter. These processes are also affected by the energy E from the X-ray quanta (similar to U_R in the generation of X-rays) and the atomic number Z, i.e., by the bonding energy of the electrons of all atoms in the irradiated matter.

In radiological diagnostics, this matter can include human tissue, bone, etc., and the respective replacement material (e.g., phantoms) with effective atomic numbers of Z = 7 and 8, as well as technical materials with higher, integral atomic numbers such as aluminum, copper, tungsten, lead, etc.

- Atomic number Z of the atoms is by definition always a whole number.
  In the case of substances consisting of more than one atom type, effective atomic number Z_{eff} is needed during irradiation of these substances to compare the interactive properties with those of air or tissue. As a result of different weighting factors, effective atomic numbers may also occur as decimal fractions.

Figure 3.6 shows mass attenuation coefficients μ/ρ (cm^2/g) of several body substances and technical materials. They are compared as a function of their corresponding density ρ (g/cm^3) and atomic number Z_{eff}, which result from different energies E during the interaction of the X-ray quanta with these materials.

- The values show, among other things, that water and Plexiglas are particularly suitable as tissue-equivalent materials (substances) for phantoms in the med-range of diagnostic X-ray energy.

Matter	Air	Fat	Plexiglas	Water	Muscle	Bone	Beryllium Be	Aluminium Al	Iron Fe	Copper Cu	Molybden. Mo	Lead Pb
Z_{eff}	7.64	5.92	7	7.42	7.42	13.8	4	13	26	29	42	82
ρ (g/cm^3)	0.001	0.92	1.18	1.0	1.05	1.92	1.84	2.69	7.87	8.96	10.21	11.34
E (keV)	Mass attenuation coefficients μ/ρ											
15	1.55	1.50	1.06	1.60	1.60	8.90	0.30	7.80	56.5	73.3	28.0	111
20	0.76	0.76	0.55	0.77	0.80	3.90	0.22	3.40	25.4	33.5	13.0	85.2
30	0.35	0.33	0.30	0.36	0.36	1.30	0.18	1.10	8.10	10.7	28.0	29.9
40	0.24	0.24	0.24	0.26	0.26	0.60	0.16	0.56	3.59	4.80	13.0	14.1
60	0.18	0.18	0.19	0.20	0.20	0.30	0.15	0.27	1.19	1.58	4.25	4.90
80	0.16	0.16	0.17	0.18	0.18	0.20	0.14	0.20	0.59	0.75	1.90	2.40
100	0.15	0.15	0.16	0.17	0.17	0.18	0.13	0.16	0.36	0.45	1.00	5.30

Figure 3.6 Attenuation coefficients and their energy dependence for different materials relevant for radiological diagnostics. Quanta energy E is similar to tube voltage U_R.

- Absorption is extremely low in the case of beryllium. Therefore, it is optimally suited for use in X-ray tube windows.

The absorption of X-ray quanta by materials such as aluminum, copper or lead is shown in Figure 3.7. This yields targeted applications:

- Filters for X-ray diagnostics are usually made of aluminum (Al). They are used to improve radiation quality. Copper filters are preferred for X-ray radiation with higher quantum energies (see section 4.1).
- With its high absorption properties, lead (Pb) provides excellent radiation protection. Radiation protection clothing typically has a lead equivalent of 0.25 to 0.5 PB (see section 4.1).

The monochromatic 60-kV X-ray radiation corresponds to a certain dose rate D. The proportion of the increase or decrease of the dose rate is shown as a percentage as a function of the material thickness (Fig. 3.7). In the case of a material thickness of 2 mm, lead absorbs all X-ray quanta and is therefore ideal for radiation protection. Aluminum and copper are used predominantly for beam filters.

- Monochromatic radiation contains only X-ray quanta with equal quantum energy (the same wavelength). In this example, 60-keV quanta are similar to a tube voltage U_R of 60 kV. Dose rate D of monochromatic X-ray radiation decreases according to the absorption coefficients of the filter material used in each case with increasing atomic numbers.
- The dose rate is the radiation dose or quantum amount applied for a certain time period to a matter (for dose information, see chapter 5).

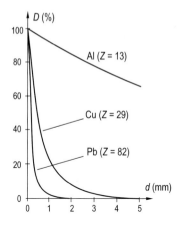

Figure 3.7 Example of absorption or reduction of X-ray quanta in a flat material layer with thickness d of aluminum (Al), copper (Cu) or lead (Pb) based on monochromatic 60 kV radiation

Based on the distribution of the absorption coefficients in Fig. 3.8 ($\tau/\rho + \sigma/\rho = \mu/\rho$), it follows for practical radiography that bones with a higher contrast are visualized as tissue in the range of a tube voltage of approx. $U_R = 50$ kV (corresponds to radiation energy E in keV). (For information on contrast, see section 9.3.) This radiographic voltage range is therefore preferred for bone radiography.

In the range of higher radiographic voltages U_R, resulting in higher X-ray radiation energy amounts, bones and tissue are attenuated in an almost identical manner. They produce only a low contrast on the X-ray image, i.e., the boundaries between tissue and bone are unclear. High tube voltages of U_R starting at approx. 125 kV are used, for example, in chest radiography to "irradiate" the ribs.

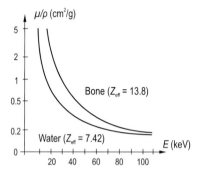

Figure 3.8 Mass attenuation coefficient μ/ρ (cm²/g) of water (corresponds largely to human tissue with respect to its absorption behavior) and bones as a function of the radiation energy. As radiation energy E increases, mass attenuation coefficients μ/ρ of bones and water (tissue) also increase accordingly.

3.3 Hard and Soft X-rays

To produce a good X-ray image, it is important to always provide the correct quantum energy or dose rate for the object to be visualized. Every object has its own, narrow, optimum kV range. Exceptions to this are alternative techniques, e.g., the high-kV technique. For chest radiography, this technique uses tube voltages U_R of typically more than 120 kV with the objective of achieving bone transparency at the lowest radiation exposure and for the shortest exposure time.

In the range of tube voltages U_R (kV) used for practical radiography and the resulting quantum energies E (keV), there are two different viewing methods when X-ray quanta penetrate matter.

- Absorption decreases in the case of the "high-kV technique" (Fig. 3.9-A), with tube voltages U_R starting from approx. 110 kV and higher-energy X-rays (quanta). Scatter increases but is diverted mainly in the direction of the primary radiation so that the majority of the X-ray quanta impinge on the image receptor.
- In the case of tube voltages U_R < 110 kV, the terms "soft" X-rays or "low-kV exposure technique" are used (Fig. 3.9-B). Absorption predominates in the case of low-energy X-rays. Scatter occurs in all directions so that the majority of the X-ray quanta do not impinge on the image receptor.

Soft tissues and bones are penetrated by hard beams and blackening occurs in all regions of the X-ray image. However, due to the scatter occurring mainly in the direction of the primary radiation, the light-dark difference is reduced, i.e., as the tube voltage increases, the contrast decreases (the proportion of scatter radiation "fogs up" the exposure).

The advantages of the high-kV exposure technique are:

- Dose reduction,
- Shorter exposure times and therefore

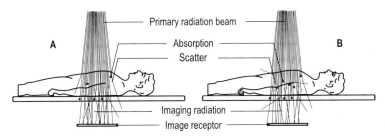

Figure 3.9 Schematic diagram of the absorption and scatter for "hard" radiation (A) and "soft" radiation (B)

- Reduction in motion unsharpness
- Blurring of absorption differences.

The reduction in the object density range (object contrast, radiation contrast, see chapter 9) can be viewed as advantageous since the increasing tube voltage results in a matching of the absorption differences with respect to atomic number Z of the matter to be penetrated; e.g., lungs, bones, and soft tissues are visualized for a usable diagnosis.

High-kV exposures are produced for

- Chest diagnosis
- Gastro-intestinal diagnosis and
- Spinal exposures.

The "low-kV technique" is used when it is preferable to work with low voltages, e.g., for mammography at U_R approx. 25-32 kV.

The advantages of the low-kV technique are:

- Large object density range with
- High image contrast.

The disadvantages are:

- High dose,
- High exposure times and therefore
- Increased motion unsharpness.

3.4 Absorption Edges

The dependence of the absorption coefficients on the energy of the X-ray quanta is a constant curve for all materials in wide ranges. However, for several keV values, noticeable "jumps", known as edges, occur in the course of the constant curve. These are referred to as "absorption edges."

In radiological diagnostics, the curve (the position) of the absorption edges of several materials is used for certain exposure techniques, e.g.:

- In vascular examinations in angiography using iodine as the contrast agent to achieve a high "iodine contrast" (see section 9.3).
- Another application involves changing the spectral composition of the X-ray radiation using "edge filters."

The absorption edge of iodine is 33.2 keV (Fig. 3.10-A). Therefore, iodine-containing contrast agents absorb X-ray quanta with energies of just over 33.2 keV particularly well (generation of the high "iodine contrast").

3.4 Absorption Edges

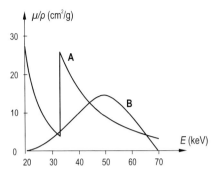

Figure 3.10 Relative spectral distribution of bremsstrahlung (B) after irradiation of a water phantom with a thickness of 15 cm and the characteristics of the mass attenuation coefficient μ/ρ (cm^2/g) of iodine (A)

These absorption edges are explained by the prescribed energy level of the electrons in the atom. X-ray quanta with higher energies than the bonding energy of the K electrons of the irradiated matter, may interact with these, i.e., transmit their energy to the K electrons.

X-ray quanta with energy lower than the bonding energy of the K electrons in the irradiated matter cannot interact with these electrons. As a result, they have less of an opportunity to emit their energy, i.e., to be absorbed. This explains the considerable difference in the value of the absorption coefficient above and below the K energy level.

As a result, the characteristic absorption edges for all materials (matter) is a function of the atomic number Z and is based, consequently, on the relevant bonding energy E of the electrons around the atomic nucleus or ΔE (K shell to L or M shell, etc.)

3.4.1 Iodine contrast

Vessels are not visible in X-ray images because their absorption coefficient does not differ from that of their surroundings. Prior to vascular examinations (angiographies), iodine-containing contrast agents are therefore typically injected into the vessels. They absorb the X-rays (quanta) better and consequently generate a contrast with respect to the vessel surroundings.

In the case of vascular examinations using iodine-containing contrast agents, tube voltages U_R of preferably 63 kV and, ideally, not more than 70 kV are used (depending on the patient's condition). As a result, the intensity of the imaging radiation is high to the right of the absorption edge in the region of the high iodine contrast (Fig. 3.10). This leads to high-contrast visualization of the vessels, using contrast agent to visualize the respective vessels.

3.4.2 Edge filter

An interesting use of absorption edges is to produce filters, known as edge filters, from materials with an absorption edge for a certain application at an appropriate energy (bonding energy E or ΔE of the electrons). Using such edge filters, the spectral composition of X-ray radiation may be modified in a targeted manner.

The absorption edge of molybdenum is approx. 21 keV (Fig. 3.11-A). The gray regions (Fig. 3.11-B) show the spectrum before the molybdenum edge filter (normalized visualization). When using the edge filter, the X-ray image is visualized largely by the characteristic radiation (white region): in the case of molybdenum, with $Z = 42$ for K_α line = 17.5 keV and K_β line = 19.65 keV.

Figure 3.11 shows the dependence of mass attenuation coefficient μ/ρ (cm^2/g) of a molybdenum edge filter (A) and its effect on the spectrum of an X-ray tube with molybdenum anode (B) for mammography. An edge filter of molybdenum ($Z = 42$) with a thickness of approx. 30 µm attenuates both the hard and the soft X-rays within the continuous radiation spectrum.

The hard X-rays within the continuous radiation spectrum generate insufficient contrast and the soft X-rays are absorbed within the irradiated object (by the matter or absorber) and do not reach the image receptor. For this reason, the characteristic radiation of the K_α and K_β lines desired for mammography imaging are absorbed only minimally.

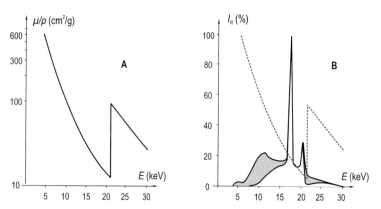

Figure 3.11 Example of an X-ray spectrum prior to and after filtering with a molybdenum edge filter, using an X-ray tube with a molybdenum anode (B)

3.5 The Distance Law

For diagnostic use, the X-ray beams are collimated near the focus so that they penetrate the object in the form of a cone of radiation and strike the image receptor. Since X-rays travel in a straight line, their intensity decreases as the distance from the focus increases and the growing surface area of the beam increases.

Figure 3.12 shows the following: dose rate D_0 (intensity of X-ray radiation) is only D_0/d^2 (level E1) at a distance of d. The dose area product with respect to surface a^2 is only $(D_0/d^2) \cdot a^2$ at a distance of d. This formula applies to all radiation levels; from the focus to the object across E1, E2 and E3.

At twice the distance from the focus, the dose is reduced by 4 with respect to the quadrupled surface – it is spread across the larger surface area. The opposite is also true. For information on dose area product, see section 5.5.

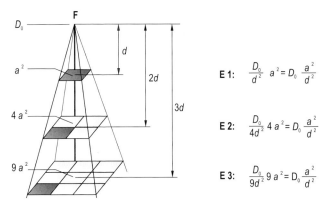

E 1: $\dfrac{D_0}{d^2} a^2 = D_0 \dfrac{a^2}{d^2}$

E 2: $\dfrac{D_0}{4d^2} 4 a^2 = D_0 \dfrac{a^2}{d^2}$

E 3: $\dfrac{D_0}{9d^2} 9 a^2 = D_0 \dfrac{a^2}{d^2}$

Figure 3.12 Quadratic dependence of the dose or dose rate starting from the center of radiation F (focus, focal point) to the object

4 The Quality of X-rays

The term "radiation quality" is generally to be understood as the average or effective quantum energy of X-ray radiation. Fundamental parameters in addition to kV, i.e., the tube voltage U_R and its voltage shape (see Fig. 4.5), are the characteristics of the filter and its half-value layer.

X-rays that are generated at a low tube voltage $U_R < 110$ kV are generally termed "soft," and X-rays generated at a high voltage U_R of approximately 110 kV "hard" (see section 3.4).

The half-value layer (HVL) thickness of X-ray radiation or the degree of homogeneity is used as a precise definition of its quality.

4.1 Half-value Layer Thickness

The half-value layer thickness, half-value layer (HVL) or half value thickness (HVT) is the thickness of a material that reduces by half the flux density of the X-ray quanta I_ψ (the number of the primary quanta of a limited ray beam) when this material is penetrated (Fig. 4.1). Similar to the half-value

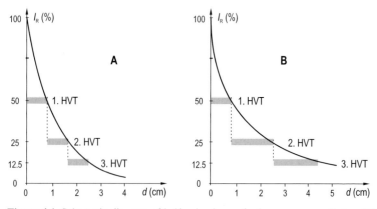

Figure 4.1 Schematic diagram of half-value layers for homogenous radiation (A) and heterogeneous radiation (B)

layer thickness, the tenth-value thickness or tenth-value layer thickness can be used to determine radiation quality. It is the thickness of a material that reduces the flux density of the X-ray quanta by 1/10 (10%), 1/100 (1%) or $1/10^n$ after penetrating a material.

Homogenous radiation as well as monoenergetic and monochromatic radiation, consists of quanta of only one type of energy. The HVL is therefore constant since homogenous radiation remains homogenous after the filter. The same HVL is measured after it is filtered, i.e., 1st HVL = 2nd HVL = 3rd HVL.

In heterogeneous, heteroenergetic and heterochromatic radiation, the X-ray spectrum contains quanta of different levels of energy. The greater absorption of the "soft" radiation mixture, the lower energy quanta, causes the secondary X-ray radiation to "harden."

4.1.1 Homogeneity

The homogeneity H as a physical parameter represents the energy distribution of the quanta (like the homogeneity factor, heterogeneity or inhomogeneity). It is calculated by the ratio of the first half-value layer (1st filter) to the second half-value layer (2nd filter), etc.

- When the radiation is homogenous, 1st HVL = 2nd HVL, i.e., the homogeneity is a constant 1 (Fig. 4.1, A).

With heterogeneous radiation, the amount of lower energy quanta (soft, long-wave radiation components) falls after each filtering. The X-rays are said to "harden." This produces a higher HVL for each additional filtering (Fig. 4.1, B).

- With heterogeneous radiation, the 2nd HVL is always higher than the 1st HVL, the 3rd is higher than the 2nd HVL, i.e., the homogeneity H is always < 1 (Fig. 4.1, B).

4.1.2 ICRU radiation quality

The intensity of an X-ray source can be measured (see section 4.4). Its quality especially depends on prefiltering, the tube voltage U_R and its waviness.

The ICRP (International Commission on Radiological Units and Measurements) is responsible for preparing the guidelines for radiation protection. It has also defined standardized radiation, the ICRU radiation quality:

- ICRU radiation quality exists when radiation has a half-value layer of 7 mm aluminum after being prefiltered through 22 mm Al (see section 4.2).

4.2 Filters

Each X-ray spectrum has low-energy bremsstrahlung components that scarcely penetrate the body due to the photoeffect.

Due to their larger wavelengths, they are chiefly absorbed by the object to be irradiated (to the disadvantage of the patient), or are scattered so strongly that they do not contribute to the image. The radiation that would harm the patient without contributing to the image is therefore filtered.

Figure 4.2 shows the following: F_{eig} = Inherent filtration (by the exit window of X-ray tube X). After 3 filters of thickness $d = 1$ mm aluminum are irradiated, the intensity I_0 is reduced to I_{Fa}, I_{Fb}, I_{Fc}, and the measured dose rate is correspondingly lower.

4.2.1 Filters to improve radiation quality

To improve the radiation quality, a part of the undesirable longer wave (lower energy) radiation is absorbed by the X-ray window of the X-ray tube. The absorption by the X-ray window is also referred to as the inherent filtration F_{eig} (Fig. 2.4). Depending on the required application of the X-ray tube, the intrinsic filtration approximately corresponds to an Al equivalent of 1.0 to 2.5 mm.

- By additionally inserting filters in the beam path, the absorption of the long-wave radiation (lower energy quanta) is promoted. The goal is to obtain homogenous radiation.

In cases where there is to be a more or less continuous transition (path) to harden the radiation, special filters are used with angled, wedge-shaped edges.

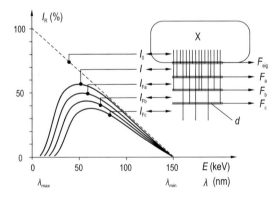

Figure 4.2 Schematic diagram of the filtering (F_{a-c}) of low-energy radiation

4.2 Filters

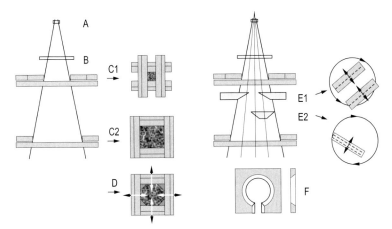

Figure 4.3 Schematic diagram of different options for beam shaping and filtering in or at the collimator (for collimators, see section 7.4)

- To compensate for the density of adjacent objects with large differences in absorption such as tissue and bone or in cardiology (vessels and tissue), wedge-shaped, semitransparent filters or slim, finger-shaped path filters are used.
- To compensate the density when imaging the pelvis, foot, shoulder, pelvic spinal column and lumbar spinal column or the skull, homogenizing filters can be inserted in the accessory rails of the collimator.

Figure 4.3 shows: inherent filtration by X-ray tubes (A); prefiltration to absorb low-energy radiation such as 1 mm Al (B); primary shaping of the beam field (C1, C2, or iris D). To compensate the density of objects with large differences in absorption, wedge-shaped, semitransparent filter diaphragms (E1), finger-shaped path filters (E2) and homogenizing filters are used; in this instance, for images of the skull (F) (see chapter 7).

4.2.2 Al equivalent

In most X-ray diagnoses, aluminum filters are used. Copper filters are primarily used for high-energy imaging (tube voltage U_R of 125 kV to 150 kV). In special cases, lead is also used.

Independent of the filter material, every material that lies in the beam path when taking X-rays is defined by the Al equivalent, also known as the hardening equivalent or equivalence.

The Al equivalent is the thickness d of the filter material that has the same absorption of low-energy X-rays (quanta, photons) as aluminum given a

4 The Quality of X-rays

specific radiation quality. The comparative value to aluminum is always used for every other material. Al equivalents (hardening equivalent) of 1 mm Al, 2 mm Al, 3 mm Al, etc.

In technical documents, the following is used to indicate the degree of filtering:

- Inherent filtration   1 mm Al at 80 kV
- Cu prefilter/hardening equivalent (Al)   0.1 mm*/3.5 mm**
- or   0.2 mm*/7.1mm**
- or   0.3 mm*/10.8 mm**

* thickness d of the copper filter
** associated mm of Al equivalent

Al equivalents are also indicated when components such as exposure measurement chambers or the patient table influence the radiation quality of the imaging, such as:

- Patient table   0.6 mm ($\pm$ 0.1) Al at 100 kV/ 2.7 mm Al HVL* (DHHS according to FDA)**
- or   0.65 mm ($\pm$ 0.1) Al at 100 kV/ 2.7 mm Al HVL* (IEC 601-1-3)

* Half-value layer, see section 4.1
** see appendix

Using an example of different tube voltages U_R, Figure 4.4 shows the effective half-value layers by indicating the Al equivalent.

4.2.3 Lead Equivalent

Like the Al equivalent, the lead equivalent is a relative value used for materials that attenuate the flux density of radiation quanta to the same degree as the comparative material at the existing thickness.

Filtration	HVL at U_R = 40 kV	HVL at U_R = 60 kV	HVL at U_R = 80 kV	HVL at U_R = 100 kV
1 mm Al	0.79	1.02	1.30	1.85
2 mm Al	1.30	1.70	2.30	3.00
3 mm Al	1.70	2.30	3.00	4.00
4 mm Al	2.00	2.60	3.60	4.80

Figure 4.4
Al equivalents and half-value layers as a function of different tube voltages U_R

The lead equivalents are chiefly used in the context of radiation protection (see chapter 5).

4.3 Tube Voltage and Ripple

The tube current I_R (mA), the tube voltage U_R (kV) and the duration of exposure (s) are necessary data for producing an X-ray image useful for diagnosis. They are called exposure data or imaging parameters (see section 9.1).

4.3.1 Tube Voltage

In diagnostic radiography, the tube voltage U_R (kV) plays an essential role in radiation quality. The higher the tube voltage, the shorter the wavelength of the radiation generated (quanta). As a result, the average radiation energy is higher and the penetration of quanta through matter is greater.

- The tube voltage U_R determines the penetration ability of the X-rays.
- When the tube voltage U_R increases, the amount of X-ray quanta penetrating the matter increases, and its absorption decreases.
- Increasing the tube voltage U_R from 70 to 77 kV, for example, decreases patient radiation exposure by about 50% due to the reduction of soft radiation.
- The tube voltage U_R has a major effect on the X-ray image (see section 9.3).
- The lower U_R, the higher the long-wave soft radiation, which is only used for special types of investigation (such as mammography) and represents a higher dose load for patients.

4.3.2 Ripple

The quality of X-rays is substantially influenced by its ripple, in addition to its tube voltage U_R (kV).

The lower the ripple or waviness W, the better the dose yield (Fig. 4.5) and radiation quality. Absolute direct voltage is ideal: the cathode electrodes are accelerated in a near-constant voltage potential U_R and generate a diagnostically valuable X-ray spectrum.

The different voltage waveforms strongly influence the diagnostically relevant dose yield and hence the radiation quality.

With half-wave and two-pulse generators, the tube voltage U_R drops to 0 potential at each half-wave, i.e., the amount of low-energy cathode electrons

4 The Quality of X-rays

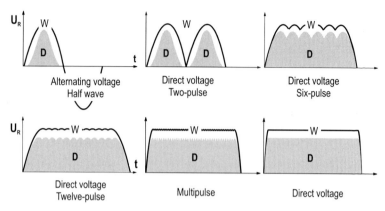

Figure 4.5 Schematic diagram of high voltage waveforms U_R, waviness W and dose yield (gray area D)

is very high and therefore comprises a larger portion of low-energy X-ray quanta.

At a tube voltage of $U_R = 100$ kV, for example, a 100 kV bremsstrahlung spectrum occurs only at the instant of the voltage peak (U_{Rmax}). A 50 kV spectrum occurs while the voltage, for example, passes through the value of $U_R = 50$ kV.

If during a sine period several rectified half-waves occur (phase shift), it is referred to as a 6-pulse or 12-pulse voltage. Operating capacitors smooth the decreasing voltage components between the phases to generate diagnostically useful direct current.

The rectified tube voltages U_R that depend on the generator type always have a certain amount of residual ripple. The ripple is a measure of the quality of the tube voltage U_R that is expressed in percent (%).

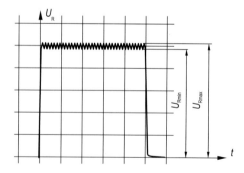

Figure 4.6
Ripple (W) of a multipulse generator.

The ripple *W* of a 6-pulse generator is approximately 13%. The waviness of multipulse or converter generators is comparable to that of 12-pulse generators at approximately 3%.

$W = ((U_{R\,max} - U_{R\,min})/U_{R\,max}) \cdot 100\%$

Advancing technology (semiconductor and microprocessor engineering, etc.) has eliminated the complex design of older generations of generators, and has led to what is known as multipulse or converter generators (see section 7.2).

4.4 Measuring X-Rays

X-rays cause ionization in air or gases. This causes changes in the electrical conductivity of the material to be irradiated. This property is exploited in automated exposure timers, radiometers and dosimetry chambers such as the Diamentor or Caremax measuring devices.

4.4.1 Ionization

As mentioned previously, if an non-decelerated X-ray quantum contacts a shell electron of an atom of the medium (the measurement chamber in this instance), this electron can be bumped out of its shell when the quantum transmits to the shell electron at least the energy that corresponds to its binding energy to the nucleus of the medium of the measurement chamber. This process is termed ionization. The released electron then moves freely within the material.

In Figure 2.6, section 2.1.3, this process is schematically represented with the example of X-ray generation. 2b is the X-ray quantum, and 5 is the released shell electron of a gas molecule.

4.4.2 Measuring Procedure

Figure 4.7 shows the two electrodes of the measurement chamber (+ and −). They are charged by a current source (U) and form a field charged from positive to negative (like a capacitor). X-ray quanta (X) penetrate the measurement chamber shown here.

When a high-energy X-ray quantum impacts the gas molecule (M), an electron can be released from the atomic shell. The ionized, positively charged molecule migrates in the electrical field toward the negative electrode. The negatively charged electrons released during ionization migrate to the positive electrode. More or fewer charge carriers are generated in this gas

4 The Quality of X-rays

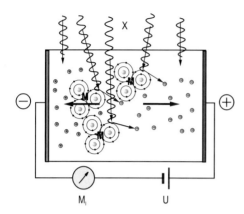

Figure 4.7 Schematic diagram of an ionization measurement chamber

depending on the intensity of the X-ray radiation. The changes in charge from gas ionization (with a flowing ionization current) are demonstrated during discharging by a measuring instrument (M_1).

By measuring ionization current, you can determine the intensity of the radiation (dose rate). Only the ionization is measured, i.e., the ion dose is defined by the quotients of the generated positive or negative electrical charge (dQ) (the positive or negative charge carrier). These are formed by the radiation in the air per unit mass (dm): $I = dQ/dm$.

- The unit of the mass m is kg (kilogram).
- The unit for the electrical charge Q is the Coulomb (C).
- The SI unit for the ion dose is therefore: $I = dQ/dm$ (C/kg)

The current generated in the measurement chamber (M_1) is proportional to the dose rate (for information on dose rate, see section 5.4). For other options for measuring the dose rate, see "Automatic exposure controls" in chapter 9.

5 Dose

Despite the successful introduction of non-ionizing imaging systems such as ultrasound, MRI or endoscopic examinations in the clinical routine, the average exposure to radiation among the population has increased during the past several years (due in part to radiological medical applications).

This chapter deals with the significance of radiation exposure for humans. The typical dose terms used in radiology are explained and measures are described to keep radiation exposure at acceptable levels.

- "An X-ray examination is justified if the patient obtains a significant benefit from the radiodiagnostics and the radiation risk is estimated to be minimal."

5.1 Radiation Exposure

Radiation exposure refers to the process by which a person receives a body dose, regardless of the type of radiation, e.g., natural or medical.

Radiotherapy uses the biological effect of radiation (i.e., the irradiation of living things) in a therapeutic manner. In diagnostic applications, this is an undesirable side effect. The biological effect of radiation leads to somatic damage (changes in the structures and functions of the irradiated organism outside the germ cell) as well as genetic damage (including natural radiation, changes to the genes) and, therefore, must be kept as small as possible.

According to the Federal Agency for Radiation Protection (1994), the proportions of the average annual exposure to radiation with a average effective dose in mSv (Sv = sievert; see appendix) are as follows:

- Medicine: approx. 1.50 to 2.00
- Respiratory air: approx. 1.35
- Terrestrial radiation: approx. 0.42
- Cosmic radiation: approx. 0.35
- Nutrition: approx. 0.25
- Other: approx. 0.06

5 Dose

5.1.1 Natural radiation

Natural radiation is composed of:

- External exposure to radiation, e.g., from the surroundings or the environment (terrestrial radiation) such as ionizing radiation from rocks, building materials, bodies of water, the human body itself, but also from cosmic radiation and
- Internal exposure to radiation (artificial radiation, factors associated with civilization), e.g., from the respiratory air, food, radiation applications in research and technology, nuclear facilities, nuclear weapons tests.

The annual dose of natural radiation is approx. 2.4 mSv. This value is used as a comparative measure for other exposures to radiation.

Adding the average medical exposure to radiation from diagnostics such as radiodiagnostics, radiotherapy or nuclear medicine yields an average annual load of about 4.00 to 4.50 mSv per person.

5.1.2 Medical exposure to radiation

The average exposure to radiation in the Federal Republic of Germany is significantly higher than in other European countries. Most of the radiation exposure associated with civilization is due to medical imaging with ionizing radiation. For this reason, from the perspective of radiation hygiene, this area requires a high degree of sensitivity.

Figure 5.1 compares the relative frequency of radiodiagnostic examinations arranged in groups (including nuclear medicine) with the proportion of the effective dose within each group.

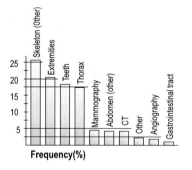
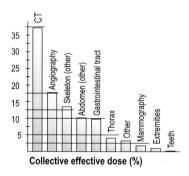

Figure 5.1 Percentages of radiodiagnostic examinations and percentages of the effective dose in radiodiagnostic examinations (source: Federal Agency for Radiation Protection)

5.2 Physical Dose Quantities

The dose measures the effect of radiation. If ionizing radiation interacts with matter, it gives off at least part of its energy. For example, radiation that is absorbed by the body (object or matter) ceases to exist; its energy is, thereby, passed onto the body (object or matter).

- As a result, the dose is defined as the radiation energy that is transferred by absorption to a certain quantity of matter.

The different dose terms are discussed and defined by international organizations such as the ICRU (International Commission on Radiological Units and Measurements), ICRP (International Commission on Radiological Protection) or the WHO (World Health Organization).

In dosimetry and radiation protection, the following dose terms and SI units (SI = Système International d'Unités) are used.

5.2.1 Effective dose E

In order to evaluate and compare different exposures to radiation the term "effective dose" was introduced. It is a measure of the risk assumed by a person who is subjected to the action of ionizing radiation. The effective dose evaluates the risk for the occurrence of possible stochastic effects in the exposure of individual organs and tissues or the entire body.

The effects of radiation can be classified as follows:

- Deterministic effects:
  Causal effects that necessarily occur at an exposure above certain dose threshold values, i.e., the severity of radiation damage is a function of the dose. Below a certain dose threshold value clinical symptoms do not occur.
- Stochastic effects:
  Pseudo-statistical effects that occur with a certain probability after a longer latency period; latency period = asymptomatic phase between the action of ionizing radiation on an organism and the occurrence of recognizable symptoms.
- Genetic effects:
  Radiation effects that cause a change in the genotype.

The effective dose cannot be measured directly. It must be calculated based on the different organ doses, which in turn depend on the average energy doses and the so-called radiation weighting factors (see section 5.3).

5.2.2 Energy dose D

This is the dosimetric basic quantity that is physically, clearly definable and is valid without limitation. The absorbed energy is the average energy given off by ionizing radiation in matter, e.g., in biological tissue.

The energy dose D (formerly known as "rad" from the English "radiation absorbed dose") is the radiation energy dE_{abs} locally absorbed by a medium of density ρ divided by the mass m of the irradiated volume element dV.

$$D = \frac{\text{Absorbed energy}}{\text{Mass}} = \frac{dE_{abs}}{dV \cdot \rho} = \frac{dE_{abs} \, (J)}{dm \, (kg)} \, (Gy)$$

The unit of energy is the joule (J); the unit of mass is the kilogram (kg). Therefore, the SI unit for the energy dose D was established as joule per kilogram. The unit 1 J/kg is designated 1 Gy (gray).

- As a result, a dose quantity measured in gray (Gy) indicates an energy dose (without consideration of the tissue weighting factors (see section 5.3)).

The energy dose cannot be measured directly in the body (i.e., tissues, organs, etc.) To determine the energy dose in a given material (e.g., organs, tissues, etc.), the energy dose in air must be multiplied by the ratio of the corresponding mass energy absorption coefficients for the medium and for air (see section 3.3 for absorption coefficient). When reporting energy doses the absorber material (medium) must always be specified.

In addition to the elementary physical dose quantity of the energy dose, the ion dose and the kerma are also used. These more nearly conform to the metrological or computational requirements.

5.2.3 Ion dose J

The term, ion dose, was introduced to obtain metrological confirmation of the intensity of X-ray radiation (see section 4.3).

It is defined as the electrical charge dQ created through the irradiation of a volume of air divided by the mass m of the irradiated volume element dV with the density ρ. The unit of measurement is coulomb per kilogram (C/kg).

$$J = \frac{\text{Electrical charge}}{\text{Mass}} = \frac{dQ}{dV \cdot \tilde{n}} = \frac{dQ \, (C)}{dm \, (kg)}$$

5.2.4 Kerma K

Kerma (kinetic energy released in matter) is the ratio of the sum of the initial kinetic energies of all charged particles (dE_{tran}), which are released in a volume element of the irradiated material through indirect ionizing radiation divided by the mass m of the irradiated volume element dV with the density ρ. Like the energy dose, the unit of measurement is the gray: J/kg = Gy.

$$K = \frac{\text{Absorbed energy}}{\text{Mass}} = \frac{dE_{trans}}{dV \cdot \tilde{n}} = \frac{dE_{trans} \, (J)}{dm \, (kg)} \, (Gy)$$

As in the case of the energy dose, the kerma must also always specify the absorber material (medium).

5.3 Dose Quantities in Radiation Protection

The dose quantities described below are defined for special applications in radiation protection. They can be split into two categories:

- Dose quantities such as the equivalent dose, local dose and personal dose.
- Body dose quantities such as the organ dose and the effective dose.

5.3.1 Equivalent dose H

At an identical energy dose, the different types of radiation, e.g., X-rays, alpha or neutron radiation, exert a biological effect of varying intensity in the body tissues. This means that the energy dose by itself does not adequately describe the biological effect of radiation in the human body. Therefore, the energy dose is defined more precisely at one point in the tissue with the assistance of *quality factors Q*.

According to a recommendation of the ICRP, in the future the term *"radiation weighting factor"* should be used instead of the quality factor and have the same meaning. The numerical values of Q or the radiation weighting factors already consider the varying biological effectiveness, i.e., the different contribution of the individual tissues or organs to the radiation risk. They are defined by agreement.

The energy dose, which includes the biological effect of radiation, is designated the equivalent dose H and is specified in sieverts (Sv). It is multiplied by the tissue weighting factors (quality or also rating factors) of the affected organs, tissues, etc. Quality factors Q have no dimension; they are a measure of the biological effect of radiation.

5 Dose

For the energy range of radiodiagnostics the tissue weighting factor (quality or also rating factor) Q is nearly 1. Therefore, in this case the amount of the equivalent dose H is approximately the same as the energy dose D. Like the energy dose, the SI unit for the equivalent dose is joule per kilogram. However, the unit J/kg is specified in sieverts (Sv) for the equivalent dose.

- Therefore, a dose quantity designated in sieverts (Sv) indicates that a dose is meant which takes into account the tissue weighting factor (quality or also rating factor) Q.

Equivalent dose of the skin (skin dose)

The skin dose, or unit skin dose, erythema dose (erythema, from the Greek erythema meaning redness, reddening; redness of the skin) is an outdated measure of the dose in radiobiology, which was based on the erythema as a radiation-caused skin reaction. In radiation protection it is the equivalent dose of the skin (energy dose D multiplied by the tissue weighting factor, that is, the quality or rating factor Q).

5.3.2 Local dose

The local dose is defined as the measured "equivalent dose for soft tissues" at a certain location. It is used, for example, to assign rooms in a hospital to different "radiation protection areas" according to the annually expected local dose (see section 5.7).

5.3.3 Personal dose

The personal dose is the "equivalent dose for soft tissues" measured at a representative location, e.g., with film dosimeters at the surface of the body.

5.3.4 Body dose

The body dose is a collective term for the partial body dose and the effective dose. It is usually calculated from the (measured) personal dose. It is the equivalent dose that is actually absorbed in the body or by the skin. It is averaged over the critical volume of the body or, if expressed in terms of the skin, averaged over the critical surface area of the skin.

The critical volume is the volume of a certain organ, tissue or body part. In case of an unequal dose distribution in the body it is relevant for calculating the body dose. If the body dose in a certain organ is meant, one speaks of the "bone marrow dose" or the "gonad dose."

Radiation protection measures must be calculated in such a manner that "persons who are occupationally exposed to radiation" will not exceed the prescribed threshold value for the body dose during the course of an entire calendar year when properly using the X-ray system.

For whole-body or partial-body exposure the "effective dose" is calculated by multiplying the equivalent dose of each irradiated organ with the corresponding weighting factor or the quality or rating factor Q and the sum of these values.

5.4 Dose Rate

The dose rate is the radiation dose administered per unit of time, more precisely the differential quotient of the energy dose and the time. It is usually specified in terms of one hour and reported as grays per hour (Gy/h) or sieverts per hour (Sv/h). The term dose rate is applicable to all dose quantities and depends on the unit chosen for the dose.

- Dose rates are the differential quotients of the specific doses according to time.
- Since the electricity produced in the measurement chamber is proportional to the dose rate (see chapter 4.3), the dose can be calculated by multiplying the dose rate with the irradiation time.

Example: At a constant dose rate of 0.6 µGy/s and an irradiation time of 3 minutes a dose of 0.6 µGy/s · 180 s = 108 µGy is applied at the image intensifier or the flat detector input.

When putting equipment into operation the dose rate at the input of an image intensifier or flat detector is adjusted to fixed values according to a specified test procedure. The internationally required value for an X-ray image intensifier with an input format of 22 cm is set at 0.54 µGy/s.

Similar to the named dose terms, the dose rates distinguish between

- Ion dose rate
- Energy dose rate
- Equivalent dose rate
- Local dose rate

5.4.1 Ion dose rate

It is defined as the ion dose per unit of time, namely C/kg·s. Since one coulomb per second is the same as one ampere (A), the unit of the ion dose rate is also specified as A/kg.

5 Dose

5.4.2 Energy dose rate

It is defined as the energy dose per unit of time, namely J/kg·s. Since one Joule per second is the same as one Watt (W), W/kg is also used as the unit of the energy dose.

However, in practice the unit that is used for the energy dose is the established designation of gray. Therefore, the unit for the energy dose rate is Gy/s.

- A dose rate quantity measured in units of gray (Gy) indicates an energy dose rate.

5.4.3 Equivalent dose rate

It is the equivalent dose per unit of time, namely J/kg·s. One Joule per second is the same as one watt (as in the case of the energy dose). However, since the equivalent dose is a dose that considers the tissue weighting factor Q (quality or also rating factor), the unit of J/kg for the equivalent dose is specified in sieverts (Sv) and the equivalent dose rate in sieverts per second (Sv/s).

- As a result, a dose rate quantity with the designation of sievert (Sv) indicates that a dose rate is meant which takes into account the tissue weighting factor (quality or also rating factor) Q.

5.5 Dose Area Product (DAP)

The product of the area of the useful beam perpendicular to the central ray (at each location between the focus and object, Figure 5.2 on the left) and the dose is called the dose area product. It is an important value for evaluating the radiation dose of the patient. In most cases, the dose is already detected

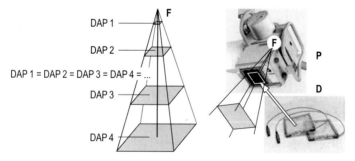

Figure 5.2 The dose area product includes variable data such as the tube voltage and tube current, examination time, filtering and field size.

at the exit from the collimator (P) with a dose measurement chamber (D) and is calculated with the assistance of a computer in the dimensions of mGy·cm². Measurement chambers for the dose area product can be inserted into the accessory rails of the collimator or can be integrated into the collimator itself.

The dose area product registers the total incident radiation experienced by the patient. It can be measured without disturbing the progress of the examination, even in the case of complex examinations. To be able to compare the dose area product with the radiation exposures of other modalities conversion into the effective dose is suitable with slight limitations. However, in complex examinations with fluoroscopy a certain uncertainty in the conversion must be considered because of the changing geometry.

5.6 SI Units

To evaluate the dose and dose rate information in older publications, it is sometimes necessary to convert it into SI units.

Figure 5.3 shows the relationship between the valid SI units and the dose units no longer approved for use since 1985/86. The activity is also mentioned here for the sake of completeness. The activity of a radioactive sample is the statistical expectation value of the quotient comprised of the number of radioactive samples in which these transformations occur.

	SI unit	Old unit	Relationship	
Energy dose	Gray (Gy) 1 Gy = 1 J/kg	Rad (rd)	1 rd = 0.01 Gy	1 Gy = 100 rd
Equivalent dose (effective dose)	Sievert (Sv) 1 Sv = 1 J/kg	Rem (rem)	1 rem = 0.01 Sv	1 Sv = 100 rem
Ion dose	Coulomb pro Kg (C/kg)	Röntgen (R)	1 R = 2.58 · 10^{-4} C/kg	1 C/kg = 3876 R
Activity	Becquerel (Bq) 1 Bq = 1/s	Curie (Ci)	1 Ci = 3.7 · 10^{10} Bq	1 Bq = 0.27 · 10^{-10} Ci

Figure 5.3 Conversion of earlier dose units into the currently valid SI units

5.7 Applied Radiation Protection

The danger of a harmful effect on tissues and organs – which can occur in medical radiodiagnostics – is extremely minimal when using modern X-ray systems and their image-generating systems with improved quality assurance and radiation protection equipment and components. To mention only a few important measures, the following lists the different possibilities for reducing the exposure to radiation:

- System-independent equipment,
  e.g., radiation protection walls, direct radiation protection at the device such as lead strips, upper and lower body protection, etc., radiation protection clothing such as coats, gloves and goggles made of lead or lead aprons, gonad protection, etc.
- System-related equipment,
  e.g., collimation near the focus and the film, semi-transparent diaphragms, Al and Cu filters, object-related additional filters, favorable exposure geometry, grids, intensifying screens, etc.
- Electronic equipment,
  e.g., RF generators, automatic exposure controls, dose rate control, selective dominance measurement, electronic collimation (without radiation), pulsed fluoroscopy, etc.
- Use of digital radiography such as Last Image Hold and a variety of image processing parameters (see section 9.2).

Through an increase in the frequency of radiological examinations as well as the number of interventional measures, the individual patient – expressed in terms of the average for the entire population – experiences a higher effective exposure to radiation. Depending on the diagnostically required X-ray examination, this can fluctuate greatly. Therefore, the medical exposure to radiation in conventional radiography techniques, fluoroscopy with or without intervention, CT examinations or imaging techniques in the field of dental diagnostics varies greatly.

- Exposure to radiation can also be reduced by having the examining physician carefully consider whether the benefit of the X-ray examination that he or she has selected for the patient is significantly greater than the expected radiation risk.

To ensure that persons who regularly or occasionally come in contact with X-rays are protected from the hazards of ionizing radiation, national and international laws and regulations have been issued and standards have been established. In Germany these include the Atomic Energy Act, the Radiation Protection Ordinance (StrlSchV) and the X-ray Ordinance (RöV). The latter is an ordinance concerning the protection from damage caused by X-rays and also applies to radiodiagnostics and radiotherapy. Unlike any other area of the law, the German atomic energy and radiation protection laws are affected by international law.

Apart from the "Dose quantities in radiation protection" (chapter 5.3), which are generally classified as belonging to radiation protection, the following section contains a series of additional radiation protection measures and defines the protection used in practice for persons exposed to radiation.

5.7 Applied Radiation Protection

5.7.1 Structural radiation protection, radiation protection plan

Before installing a radiography system, a radiation protection plan must be prepared. In this radiation protection plan the radiation protection areas are distinguished as follows:

- non-operational supervised areas,
- operational supervised areas,
- controlled areas, and
- exclusion areas.

These are all rooms in which working with ionizing radiation exceeds a certain local dose (or can be exceeded) or in which persons who remain in these rooms are permitted to receive certain body doses (Fig. 5.4). An expert evaluates the correctness of the plan.

- The non-operational supervised area directly borders the operational supervised area. This includes paths, streets, green spaces and neighboring living spaces or offices.
- The operational supervised area permits free access and a long-term stay. These are the rooms for physicians, personnel and patients. In this area persons are permitted to receive an effective dose of more than 1 mSv in a given calendar year; in terms of certain organs or superficial areas of the body even more is permitted.

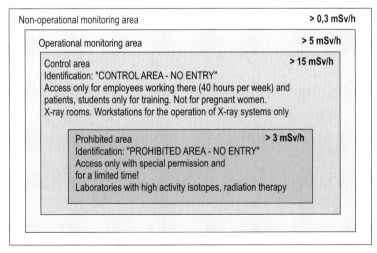

Figure 5.4 An example of classifying the radiation protection areas according to the expected local dose in certain time periods

5 Dose

- In the controlled area persons are permitted to receive an effective dose of more than 6 mSv or organ doses of 45 mSv for the eyes or 150 mSv for the skin, hands, forearms, feet and ankles in a given calendar year.
- Exclusion areas are parts of a controlled area in which the local dose rate can be higher than 3 mSv per hour.

The X-ray room itself is an additional area in which radiography systems may only be operated if the room is completely enclosed on all sides by radiation protection and it is described by the authorized expert in the permit or certificate as the room for taking diagnostic X-ray images.

Locally fixed or mobile protective screens containing lead glass or protective curtains made of flexible lead rubber are also classified as part of the structural radiation protection.

Radiation protection screens, either permanently mounted or mobile, are predominantly used in radiography systems that provide for the adjustment of system settings and/or the exposure data at a distance from the patient. This so-called remote operation of the radiography system in many cases also provides the examiner with a control room separated from the examining room by a large lead glass window.

In the case of radiography systems for general angiography, universal fluororadiography or computerized tomography, a control room is mainly equipped for the remote or secondary operation of the system, exposure and/or image processing parameters.

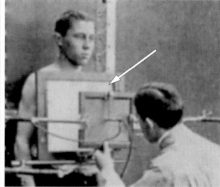

Figure 5.5 The first radiography rooms used in practice did not yet employ any radioprotective measures for either the patient or the physician. X-ray image of a knee around 1900 (left) and the thorax 1907 (right). However, this radiography room already employed a compression diaphragm (arrow), which contributed to a reduction of the scatter radiation produced in the patient.

5.7.2 Technical radiation protection

These are technical radiation protection measures on the X-ray systems themselves. They primarily have the task of protecting the examiner and the persons providing assistance against direct radiation, e.g., the scatter radiation originating from the patient, leakage radiation of the X-ray tube and other stray radiation (see Fig. 5.6).

To protect the upper body a transparent lead glass plate is attached to a carrier tripod mounted on the ceiling (Fig. 5.6-a). The cut-out circular section makes it easier to work on the patient, e.g., in interventional examinations.

A radiation protection screen with a transparent lead glass plate combined in this instance with a generator control console is positioned in front of a radiographic workstation for mammography in the examining room (Fig. 5.6 b).

For radiography systems at which the patient is mainly examined immediately, the radiation protection measures are usually directly attached to the system. Especially in the case of tiltable systems, radiation protection – corresponding to the particular movements of the system – must fulfill its function. This is achieved by means of lead rubber strips attached to the accessory rails of the patient positioning table in order to provide upper body or lower body protection at the spotfilm device (Fig. 5.6-c above). The latter is primarily used in interventional examinations on angiography systems.

Radiographic workstations with an X-ray tube beneath the patient positioning plate usually have an additional radiation protection shield, which is mounted in a fixed position (Fig. 5.6-c below).

A radiography system that meets these standards fundamentally offers the examiner and assisting personnel very good technical radiation protection. However, not only the design of the radiography system is important for the

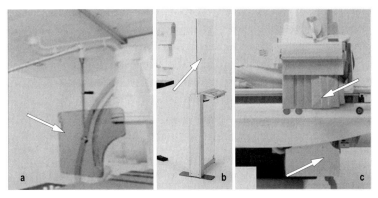

Figure 5.6 Different radiation protection devices for the examiner, currently in use

radiation dose received by the physician, assistants and the patient; the type and manner of use is also important.

In the interest of personnel, technical radiation protection should definitely be supplemented by additional measures such as the wearing of radioprotective clothing.

5.7.3 Radioprotective clothing

Specialized manufacturers offer a comprehensive collection of radioprotective clothing (Fig. 5.7) for all radiological examinations performed near the patient.

Figure 5.7 Lead rubber coat apron, cervical collar, gloves and goggles for the protection of the examiner. Ovary and gonad protection for the patient.

According to the radiological discipline, lighter coat aprons with different lead equivalent values (0.25 and 0.13 mm Pb) are preferred in surgical, orthopedic or urological applications. In longer examinations or interventions (e.g., in angiography or at cardiac catheterization labs) additional cervical collars (0.5 mm Pb) are recommended to protect the thyroid gland.

5.7.4 Radiation protection for the patient

Human beings may only be exposed to X-ray radiation within the context of medical examinations and for therapeutic purposes. The application must always occur in such a manner that the radiation dose received by the person being examined is kept as small as possible.

The International Commission on Radiological Units and Measurements (IRCP) lists the basic principles of radiation protection. According to the

IRCP, it is not possible to set particular dose values at which somatic or genetic damage occurs in humans (see section 5.1). However, the necessity of using rays on humans despite their harmful effect resulted in the compromise of establishing maximally permitted doses.

Aside from the permitted specified doses in the relevant literature (see appendix), which depend on the diagnostic questions posed and the radiological examination methods employed, there is a great variety of measures available to keep the radiation dose received by the patient as small as possible. Among others, these include the following:

- For the protection of the gonads or ovaries, the thyroid gland and the sternum: half-aprons and cervical collars made of lead rubber, or ovary protection made of lead and coated with plastic.
- The correct positioning or placement of the patient.
- The correct radiographic projection, correct selection of the focus size and the focus-to-film distance, the smallest possible collimation of the cone of radiation.
- The correctly measured components, such as the grid and film-screen systems.
- The appropriate accessories for positioning and radiography.
- The correct radiographic parameters corresponding to the radiological discipline.

For more information regarding the points above, refer to chapter 9: X-ray imaging technology.

5.7.5 Radiation dose monitoring

Radiation dose monitoring of people who are occupationally exposed to radiation and who remain or work in the controlled area requires measurement of the personal dose; this does not apply to patients. The measurements are performed with small dose measuring devices. These are the

- Pocket dosimeter and
- Film dosimeter.

They are worn on the upper body approximately at breast level; they are worn underneath protective clothing.

Pocket dosimeter

Pocket dosimeters, also rod or pen dosimeters, can be evaluated at any time to determine the received personal dose. The measurement range is approx. 0 to 2 mSv with an accuracy of approx. ±10%.

5 Dose

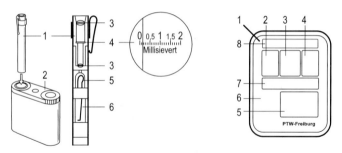

Figure 5.8
Main construction of a pocket dosimeter (left) and a film dosimeter (right)

Figure 5.8 on the left, shows the typical design of a pocket dosimeter (1) with charger (2) and display in mSv (4). In the dosimeter a quartz fiber (5) and a surrounding, insulated metal cylinder form an ionization chamber (6). When rays penetrate, the changes in the charge caused by gas ionization act upon the quartz fiber (5), which becomes slightly deformed. Optics (3) are used to display the changed position of the quartz fiber on the scale (4).

Figure 5.8 on the right, shows the main diagram for a film dosimeter. It consists of a casing made of hard plastic (1). The casing contains 5 different filters: 0.3 mm Cu (2); empty field or also empty filter (3); 0.05 mm Cu (4); 0.8 mm Pb (5); 1.2 mm Cu (6). Through the blank field (7) the impressed film number can be read; the upper field (8) is intended to be used for the name label.

Before starting to work in supervised areas subject to radiation, the pocket dosimeter must be charged to approx. 150 V with the charger. Precise measurements of the personal dose require that the measured values be read and recorded immediately after completing work.

Film dosimeter

Additionally or as an alternative, measurements of the body dose with a film dosimeter are also prescribed. Film dosimeters must be obtained from authorized monitoring agencies and must once again be returned for monthly analysis. The measurement results are recorded and are communicated in writing to the submitter or the RSO (30-year storage time period).

The measurement is performed by using what is known as the filter analytical procedure. Since different filters also absorb the X-rays (quanta) to varying degrees, the dose can be estimated from the different levels of blackening present on the X-ray film (see section 4.1 for filters). Two X-ray films with different sensitivities are inserted for a more precise estimate of the radiation dose (see section 8.1 for sensitivity).

Finger ring dosimeter

To determine the partial body dose on the hands, finger ring dosimeters – also known as thermoluminescence dosimeters (TLD) – are used. Like a stone on a decorative ring, they contain a solid radiation detector with an edge length of only a few millimeters. The radiation detector mainly consists of a material that shows the effect of thermoluminescence.

Upon irradiation, the shell electrons of these materials, e.g., lithium fluoride (LiF), are lifted to a higher, stable energy level (to what are called "traps"; see section 8.2). Through the addition of heat (approx. 200° to 300°C), these electrons are once again released from their position. Emitting light (luminescence), they return to their original state. The quantity of light that is emitted is proportional to the previously absorbed radiation energy.

5.7.6 Organizational radiation protection

Each operator of an institution in whose rooms ionizing radiation is applied must name a coworker who is responsible for the observation of all the requirements related to radiation protection in his/her area of competence. The operator of this institution can be a legal entity or a natural person. If the operator is a legal entity (e.g., the communal association of a hospital), a natural person – usually the head of administration – is named as the radiation safety officer. In medical practices, this is the owner of the practice.

Person responsible for radiation protection

The operator of an institution is the person responsible for radiation protection. This person is obligated to ensure that suitable protective measures are used to protect against radiation damage to life, health and property, and that during the operation of the radiography system the safety regulations are observed. The radiation dose experienced by persons and the general public must be kept as small as possible.

The person responsible for radiation protection is obligated to prepare suitable rooms, protective measures, devices and safety equipment for persons as well as to acquire suitable personnel and to make appropriate arrangements for the sequence of operations. This person is responsible for giving notice (registration) of type-approved radiography systems. He or she names the radiological safety officer (RSO) in writing.

Radiological safety officer

The radiological safety officer must provide proof of adequate knowledge for the task at hand. On an ongoing basis, the person responsible for radiation protection – or the RSO named by this person – must arrange to have the personal doses, the dose rates or the local doses measured and to have the

5 Dose

results recorded. He or she must, on at least a semiannual basis, provide radiation protection training for all persons working in the controlled areas. Once each year or immediately, if a prescribed dose limit is exceeded, this person must initiate an examination by the authorized radiation protection physician.

Persons exposed to radiation

All persons exposed to radiation (i.e., those persons working in the controlled area) are obligated to cooperate in the measurement of the personal dose and the examinations performed by the authorized radiation protection physician. They must also participate in the instruction provided about radiation protection.

Those individuals responsible for radiation protection must not be hindered in fulfilling their obligations or suffer any disadvantages because of their activities.

6 X-ray Systems for Diagnostics and Intervention

Modern medical imaging systems present a broad array of examination methods for diagnostics and intervention (Fig. 6.1). The technology is constantly developing, and the future holds new and unanswered questions that will continue to drive innovation. Ongoing advances in medical science, physics and technology are yielding an abundance of solutions that will continue to change, improve and complement one another. As in the past, more efficient technology will supersede less cost-effective or unproven imaging and image receptor systems. Medicine and technology will continue to develop sophisticated imaging systems with improved ergonomics and usability to protect and benefit patients and examining personnel.

The imaging systems described here are designed specifically for standard examination methods. While some systems focus on specialized and differential diagnoses, the use of one system does not necessarily preclude the use of another.

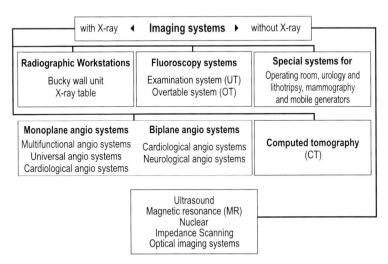

Figure 6.1 Standard diagnostic imaging systems with and without X-rays

6.1 Imaging Procedures without X-rays

The choice of imaging technique depends on the indications and preliminary diagnosis. Imaging systems that do not use radiation are discussed here in brief. For a more thorough discussion, refer to the reference literature in the appendix.

6.1.1 Sonography

Sonography (ultrasound diagnostics) is an imaging technique based on the interaction between the human body and ultrasound waves or fields. In the procedure, an ultrasound pulse is transmitted and the echo signals are recorded over time. These parameters are depicted as two-dimensional ultrasound images on a monitor (Fig. 6.2). Specialized computer programs are able to render three-dimensional images.

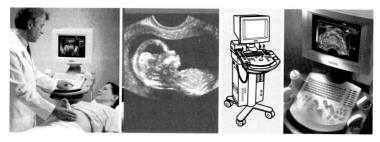

Figure 6.2 Ultrasound imaging systems, ultrasound examination and result images

6.1.2 Magnetic resonance imaging

Magnetic resonance imaging (MR or MRI) is a procedure based on the behavior of the magnetic dipoles in the nucleus of the atom (hydrogen atoms) in constant and alternating magnetic fields. Like sonography and computed tomography, MRI is classified as a sectional imaging procedure.

Every moving electrical charge can be viewed as an electrical current that generates a magnetic field. Atomic nuclei with uneven numbers of protons and/or neutrons have an intrinsic angular momentum (nuclear spin), similar to the movement of a gyroscope.

If an external magnetic field is applied (Fig. 6.3-M: main magnet), the randomly distributed nuclear spin positions in the human body align parallel or antiparallel to the direction of the field. If an electromagnetic radiofrequency field is applied at a certain angle to the aligned, vertical spin axes (nuclear spins) (Fig. 6.3-S: coils to direct the magnetic field), the nuclear spins attempt to escape the force of the magnetic field.

6.1 Imaging Procedures without X-rays

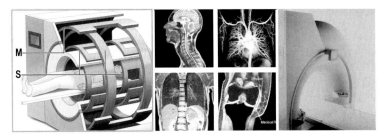

Figure 6.3 Principle of magnetic field alignment in MRI (left), result images (center) and gantry with patient table (right)

When the magnetic RF field is switched off, the nuclei return to their randomly aligned states. In doing so, they produce an electrical current in the switched-off RF coil with the same frequency as the precession frequency of the nuclear spin (movement of the gyroscopic axis). With the help of a system processor, the different signals produced by the hydrogen atoms (or molecules) are used to align the affected magnetic nuclei with the biologic tissue and produce an image; signals differ according to atomic density, concentration and bonds, as well as chemical and physical properties of the atoms.

6.1.3 Impedance scanning

Impedance scanning is a radiation-free imaging method that can be used as an adjunct to mammography (Fig. 6.4).

Patients hold a source electrode in their hand. The electrode transmits a biocompatible alternating current through the body. Sensors are used to measure the current on the surface of the breast (Fig. 6.4, arrow). Malignant tissue (from the Latin *malignus*, or bad) has different impedance characteristics, i.e., it conducts electricity differently from healthy tissue or benign lesions (from the Latin *benignus*, or good). The system processor calculates

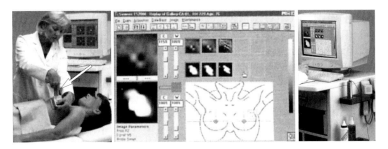

Figure 6.4 Scanning process and display on the monitor

75

6 X-ray Systems for Diagnostics and Intervention

the tissue's conductive capacity from the different data. These parameters are depicted on-screen as a two-dimensional impedance map (Fig. 6.4).

While the breast is being scanned, the probe sensors track the biocompatible voltage released by the electrode. Both impedance parameters, capacity and conductivity, are calculated by the system and displayed on the monitor as physiological real-time images or impedance images.

6.1.4 Nuclear medicine diagnostics

In terms of radiation physics, there are strong parallels between X-ray and nuclear medicine diagnostics (Fig. 6.5). In nuclear medicine imaging, however, the source of radiation is located inside the body. Radioactive markers are used inside the object for functional and localization diagnostics as well as treatment using unstable nuclides (see section 2.1.1 for information on nuclides).

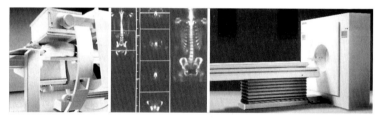

Figure 6.5 Gamma camera detector image acquisition system (left), result images (center) and whole-body positron emission tomography (PET) (right)

Although nuclear medicine is not based on X-rays, the nuclear medicine physician is nonetheless a radiologic diagnostician.

6.1.5 Optical imaging processes

Optical imaging processes complete the array of radiation-free imaging techniques. They include photography (e.g., in dermatology), microscopy (e.g., cell cultures, histology), endoscopic procedures (e.g., reflection of the larynx, stomach, colon) or ophthalmoscopy (examination of the back of the inside of the eyeball).

6.2 Imaging Procedures Using X-rays

The choice of imaging technique depends on the indications and preliminary diagnosis. Depending on the exam methods, imaging systems for radiologic diagnostics are differentiated based on the requirements profile.

The diagnostic image acquisition processes that use X-rays are discussed only briefly here. Refer to the appendix for reference materials that describe such systems in full.

6.2.1 Native image diagnosis

Native imaging (from the Latin *nativus*, or natural, unchanged) is a type of projection radiography that uses no contrast agent (also called plain film radiography).

Native images include:

- Skeletal images
- Chest images
- Mammography.

The X-ray systems used to capture such images must be equipped with image receptor components, such as a

- Film-screen system
- Imaging plate system or
- Solid-state detector system.

These systems produce single images only, not dynamic image series.

6.2.2 Examinations using contrast media

Because the differences in human-tissue density are small, it is very difficult, if not impossible, to distinguish between various organs or parts of organs. This is especially true when examining the gastrointestinal tract (stomach and colon) and visualizing blood vessels.

Examinations that use contrast media include:

- Exams of the gastrointestinal tract,
  also often studied endoscopically, except for the small intestine (until recently). Endoscopic exams of the small intestine are now also being performed.
- Cholegraphy,
  although gall bladder diagnosis mainly uses sonography (ultrasound).
- Bronchography,
  for high-resolution, detailed visualization of the bronchi.
- Arthrography,
  to visualize a joint when clinical exams and native images do not point to a definitive diagnosis.
- Myelography,
  to diagnose trauma or impingement of the spinal cord.

6 X-ray Systems for Diagnostics and Intervention

- Catheter arteriography,
  to visualize the arterial vascular system (angiography systems and spiral computed tomography).
- Venography
  or phlebography to visualize the veins.
- Lymphography
  to visualize the lymphatic vessels and nodes.
- Infusion urography
  to visualize the kidneys and urinary tract.
- Cystography, urethography
  to visualize the urinary bladder and urethra.
- Hysterosalpingography
  to visualize the uterus and fallopian tubes.

These systems, in addition to being able to produce film-screen or storage-screen images, must also be equipped with an image acquisition system, such as a

- Image intensifier or
- Flat detector.

In addition to radiographic images, sectional imaging procedures such as computed tomography (CT) are also used to visualize organs and vessels in the skull, chest and abdomen. These procedures include

- Spiral CT (to visualize organs) or
- CT angiography (CTA).

Most of the exam is computer-driven. A native CT is always performed before each contrast CT.

6.2.3 Interventional Radiology

The minimally invasive (Latin *invasio,* penetrate) interventions can replace or complement operations. The basic types are:

- Interventions of the vascular system, either arterial or venous, (arteries carry light, oxygenated blood away from the heart; veins carry dark blood towards the heart), with the help of angiography X-ray systems.
- Non-vascular (without blood vessels) interventions using fluoroscopy and multifunctional X-ray systems.

These procedures are monitored under fluoroscopy using image intensifier or flat detector systems.

6.3 Technology of X-ray Systems

For each type of exam, radiologists and radiology professionals have at their disposal a variety of radiography systems, developed as always to meet specific diagnostic and radiological requirements. They can be classified into X-ray systems for the

- Skeleton and chest
- Trauma and emergency
- Bedside radiography
- Mammography
- Dental
- Internal medicine
- Urology
- Angiography
- Surgery
- Computed tomography.

X-ray generators and tubes make up the "radiation-producing system." X-ray systems consist of stands and tables at which the patient stands or where the patient is positioned according to exam requirements. They also house the X-ray tube and the image acquisition system.

Besides direct and indirect analog imaging technology, technological progress established digital imaging technology, without which modern diagnostic imaging would be impossible.

All the X-ray systems described already employ a digital imaging chain or can be appropriately equipped (see chapter 9).

These X-ray systems can be divided into four main groups, based on their diagnostic utility as well as their historical development.

- X-ray systems that produce only individual images without prior fluoroscopy are known as X-ray imaging or radiography stations. They include skeleton and chest, trauma and emergency, bedside, mammography and dental X-ray systems.
- X-ray systems that focus on fluoroscopy or fluoro-guided images are called fluoroscopy systems. These X-ray systems are used for internal medicine and urology. They can also be used for conventional images and angiographic examinations.
- X-ray systems targeted towards angiography and surgery are among the C-arm systems even if fluoroscopy is the prerequisite for the various imaging methods, similar to the other systems discussed.

• The imaging technique for X-ray systems for computed tomography differs from the other X-ray systems. The image acquisition system consists of detector systems that convert the X-ray quanta into electrical signals. The image calculation process is exclusively digital.

6.3.1 X-ray systems for the skeleton and chest

Around 1900, tuberculosis was the most frequent cause of death in the western world. The discovery of X-rays led to the development of X-ray systems that permitted not only early detection of this disease, but also control of its course. Physicians and manufacturers therefore concentrated on thoracic exams as well as skeletal diagnostics. Because timely treatment so depend on it, early detection of lung disease, the identification of fractures and gall stones, and the localization of foreign bodies have become the primary field of activity for the radiologist.

In addition to all the standard images of standing, sitting or recumbent patients, these systems can produce special images for orthopedic, trauma, routine, emergency and chest exams as well as bedside and tabletop images.

System variations

Today, hospitals and medical practices can choose from a broad range of imaging products to equip their X-ray facilities according to their various diagnostic specializations.

Radiography systems for examining the chest and skeletal system primarily fall into the category of X-ray or radiology workstations. These systems produce native images almost exclusively. The available image acquisition systems follow:

• Film-screen and/or storage-screen systems (digital luminescence radiography, DLR) or

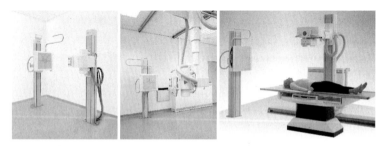

Figure 6.6 Exams of the thoracic cavity, the abdominal area, the skull and the skeletal system of standing, sitting, or recumbent patients were conducted primarily on imaging tables (right) in combination with a stand (left and center)

6.3 Technology of X-ray Systems

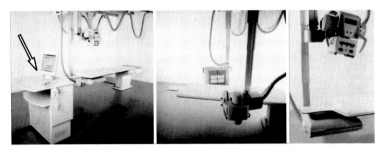

Figure 6.7 Digital flat detector radiography workstation. Imaging workstation with network connection (arrow).

- Flat detector systems.

In addition to film-screen and/or storage-screen systems, modern X-ray diagnostics increasingly uses digital imaging, such as image acquisition systems with flat detector technology. The use of imaging plates is on the decline.

For universal applications, bucky wall stands are often used as a second workstation in fluoroscopy systems.

The image acquisition system is part of what is called a bucky tray or diaphragm, which has the same capabilities as the previously described imaging systems. It also has a device for collimation of the image format near the cassette or detector, a grid for reducing scatter radiation, and a measurement chamber for automatic exposure control.

Two variations of the system are available:
- Bucky table for imaging recumbent patients, with a floor or ceiling-mounted stand for the X-ray tube and collimator.
- Bucky wall stand for imaging standing or sitting patients. Models with a fixed or tiltable bucky are available. To operate as a self-sufficient system, these models do require a wall or ceiling-mounted stand for the X-ray tube and the collimator.

Usually both radiographic workstations are combined in one exam room. This allows the tube stand of the bucky table to be used for images on the bucky wall stand. For specific lines of diagnostic inquiry, the bucky table can also be equipped with a device for conventional slice images (see section 9.4).

X-ray systems with an image receptor using flat detector technology (see Fig. 6.7) for digital imaging extend radiologic capability from a mere skeletal and chest workstation to a comprehensive functional imaging system connected to the network (see chapter 10).

6 X-ray Systems for Diagnostics and Intervention

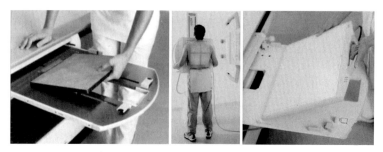

Figure 6.8 Bucky tray for cassettes (left) and flat detector (right) on the X-ray table. Center: p.a. chest X-ray at bucky wall stand with detector tray.

To ensure that older generation X-ray systems can deliver all the benefits of digital imaging in the future, flat detectors can be retrofitted as image receptors in cassette format. The advantage of these "flat-detector cassettes" is their mobility (Fig. 6.8, right). They can be inserted in the detector tray of the bucky table as well as the detector tray of the bucky wall stand in the same room (Fig. 6.8, center).

6.3.2 Trauma and emergency X-ray systems

X-ray systems for trauma and emergency exams are used mainly for imaging in the trauma room, outpatient clinic, casting room and in company medical

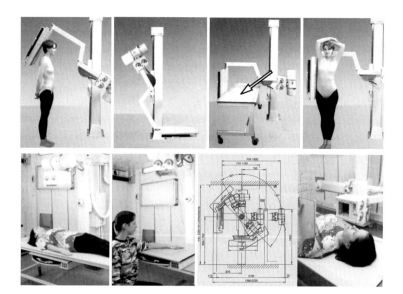

Figure 6.9 X-ray systems for routine, trauma and emergency examinations

6.3 Technology of X-ray Systems

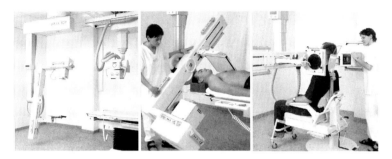

Figure 6.10 X-ray system especially for skull and skeletal imaging. Combined with an X-ray table (left and center), it becomes an radiography system for emergency, trauma and routine images.

facilities. If appropriately constructed, the X-ray imaging station (Fig. 6.9, bottom row) can also be built into mobile units, e.g., a specially designed container or bus.

The centering of the central ray on the cassette in all projections enables this radiography system to be used on standing, sitting and recumbent patients, especially to image accident victims. Patients usually lie on a specially equipped transport trolley (Fig 6.9, arrow) for care, examination and X-rays.

System variations

Using the extremely precise ORBIX technology for skull and skeletal images (the area of interest is always in the isocenter), fine adjustments of projections, reproducibility and enlargements of images are straightforward (Fig. 6.10) (see Chapter 9).

6.3.3 Mobile X-ray systems for bedside images

Mobile generators are closed, stand-alone X-ray units that can be used anywhere, usually for patients who are immobile.

Mobile X-ray generators are used in nursing and intensive care wards, operating rooms, company medical facilities, and in sports as well as veterinary medicine.

Cassette images are created on film-screen combinations or imaging plates. The cassette (with its integrated stationary grid) is positioned under the object to be imaged.

System variations

In addition to the generator with the high voltage U_R required for X-ray imaging, the mobile X-ray unit houses all the other components needed for the image.

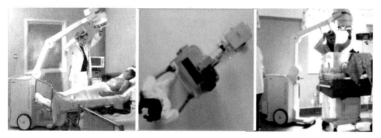

Figure 6.11 Mobile X-ray generators

A battery is recharged over a mains connection and supplies the power for X-ray generation, image processes and transport.

X-ray tubes with a collimator (usually a slot collimator) are mounted to the end of a moveable support arm so that the required projections and source-image distances can be adjusted to the position of the usually immobile patient. (See section 7.2 for more information on mobile X-ray generators.)

6.3.4 X-ray systems for mammography

For years, high expectations were placed on radiation-free imaging systems for the early detection of breast cancer in women. We know today, however, that a well-performed mammography is superior to every other exam method. Impedance scanning is one radiation-free imaging method that can be used as an adjunct to mammography (refer to the beginning of chapter 6).

Although breast tissue is highly sensitive to radiation, the radiological exam is still the method of choice for screening programs and check-ups.

Because the female breast is made up exclusively of soft tissue, differences in density are small. In addition, tissue thickness varies from the base of the breast to the mammilla. For these reasons, special mammography image systems were developed to meet the specific requirements of this examination.

The purpose of mammography is to depict the tiniest contrasts in soft tissue and microcalcifications up to 0.1 mm.

Mammography examinations are performed at imaging voltages U_R between 25 kV and 35 kV so that minor differences in density of even tiny structures are easier to recognize. To avoid producing unwanted characteristic radiation, a molybdenum anode is used to generate the X-rays. The otherwise undesirable heel effect is used to compensate for differences in thickness when generating X-rays on the anode (see section 7.3.4).

System variations

In addition to a highly accurate film-screen combination or a unit designed for digital imaging (CCD, or increasingly, the flat detector) (see chapter 8), the image receptor system also includes a grid to reduce scatter radiation and a measurement chamber for automatic exposure control. To improve the density characteristics and also reduce scatter radiation, mammography units are equipped with a highly-sensitive compression device (Fig. 6.12, arrow).

For investigating malignant processes, tissue samples need to be taken; therefore, every mammography system is equipped with a biopsy device (Fig. 6.13). (See chapter 9 for more information on stereotactic biopsy.)

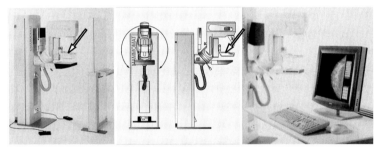

Figure 6.12 Digital mammography imaging station with flat detector. Imaging workstation with network connection (right).

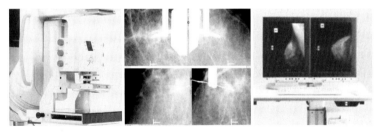

Figure 6.13 Digital mammography imaging system with stereotactic biopsy device (left). Stereo views with the biopsy system and needle marking (center). Imaging workstation with network connection (right).

6.3.5 Dental X-ray systems

X-ray images of the teeth are prepared as endoral or panorama images.

In endoral imaging, a direct technique is used to image a diseased tooth or group of teeth. A special dental film packet holds the small X-ray film in place or the patient uses his or her index finger to hold it flat against the inside of the teeth or jaw.

Panorama X-ray is also an imaging technique, although it has not come into general use. Panorama images visualize the upper and/or lower jaw. A special tube is inserted in the patient's mouth, while the X-ray film is located outside the mouth.

Figure 6.14 Dental X-ray device from 1931 (left) and a modern X-ray system (center) for the teeth, jaw and skull as well as panorama slice images (right)

To visualize all the teeth or the entire jaw region in one image, panorama tomography is widely used (Fig. 6.14). (See chapter 9 for more information on the panorama slice image).

6.3.6 X-ray systems for internal medicine

Although endoscopy is being used more frequently to address specific diagnostic concerns in the gastrointestinal tract (except the small intestine), fluoroscopy systems designed specifically for diagnosing problems in this region continue to be used with great success. Fluoroscopy systems can be enhanced to meet specific technical specifications for use in universal applications, such as

- Angiography (visualization of the arterial system)
- Phlebography (visualization of the venous system)
- Arthrography (visualization of the interior of a joint) or
- Myelography (visualization of the spinal canal).

Universal fluoroscopy systems can capture images directly and are usually equipped with an imaging system for dynamic images. For dynamic studies, an image intensifier with an X-ray television screen is attached to the spotfilm device. As flat detector technology matures, however, it will continue to replace the image intensifier in universal fluoroscopy systems.

System variations

The image receptor system or spotfilm device contains an image intensifier (solid-state detector), an assembly for cassette images (film-screen and/or imaging plate), a device for collimating the image format close to the cassette, a grid for reducing the scatter radiation, and a measurement chamber for automatic exposure. The various image receptor systems are discussed in chapter 8.

Terms such as "fluoroscopy unit", "fluoroscopy device", or "fluoroscopy system" are commonly used for "universal" radiologic diagnostics with an emphasis on internal medicine. There are two types of fluoroscopy systems: undertable (UT) and overtable (OT).

Each system must be able to create targeted images from fluoroscopy. In addition, the base system must have a motor-driven patient tabletop that moves both longitudinally and/or transversely. The table must be tiltable from the vertical to 90° Trendelenburg, as the diagnostic investigation requires. Universal fluoroscopy systems must allow the patient to be examined from head to toe without repositioning.

For dynamic studies, special X-ray devices for the OR, urology, and various angiographic exams also have an image intensifier or a solid-state detector to convert the X-ray image. System functionality is based on the needs of the particular specialty.

Undertable fluoroscopy systems

The major characteristic of undertable fluoroscopy systems is that all system movements and operations for imaging and fluoroscopy originate at the spotfilm device, i.e., the physician always works directly on the imaging system at tableside (Fig. 6.15, left).

In undertable systems, the X-ray tube is located underneath the patient tabletop. The distance between the focus and the tabletop is fixed. The X-ray tube is attached to an image receptor (spotfilm device) with a U bracket (Fig. 6.15, center) and the distance to the table can be changed as needed. In addition, the spotfilm device, whose radiation beam is limited to the size of the image, can be moved longitudinally and transversely over the patient.

To acquire images from a second plane, undertable fluoroscopy systems have a second X-ray tube with a collimator mounted on a ceiling stand (X-ray

6 X-ray Systems for Diagnostics and Intervention

Figure 6.15 Universal undertable fluoroscopy system including spotfilm device with tableside operation and image intensifier. Imaging workstation with network connection.

tube ceiling stand, Fig. 6.15, right). This enables images on a catapult bucky tray, or cassette bucky tray, which is also installed underneath the patient tabletop. A universal undertable fluoroscopy system can also be used for standard exposures.

Overtable fluoroscopy systems

The first overtable fluoroscopy systems arrived on the scene in the 1950s with the development of image intensifier television devices for medical use.

Overtable fluoroscopy systems feature remote operation. System movements, imaging and fluoroscopy are controlled either from a radiation-free control room or from behind a protective separating wall in the exam room (Fig. 6.16, top center).

The X-ray tube and the collimator are located above the patent table on a stand integrated into the basic system (system tube stand). The spotfilm device is located under the patient table and can be moved longitudinally, linked to the tube stand. One advantage over undertable fluoroscopy systems is the overtable's larger source-image distance of 115 cm (typical) or 150 cm (Fig. 6.16-8). Undertable systems are approx 65 cm. The larger distance substantially improves the image geometry (see chapter 9).

Fig. 6.16 shows a remote-controlled overtable fluoroscopy system with spotfilm device for cassette images, the remote control unit (2) with selectable image parameters and organ programs, etc. (3), and the imaging workstation (4) with network connection, and well as a mobile control panel (5) for tableside operation.

If there is no need for a second image plane, the tube stand integrated into the basic system enables oblique emissions, conventional linear or digital slice images (Fig. 6.16-7), peripheral angiographic exams (9), bedside images (6b) and/or images on a bucky wall unit (6a). For special diagnostic

6.3 Technology of X-ray Systems

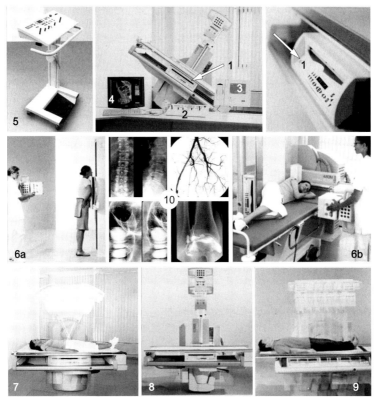

Figure 6.16 Modern universal overtable fluoroscopy system, remote and tableside operation and result images (10)

studies, an overtable system can also be equipped with a ceiling-mounted tube stand for images from a second plane.

Advances in flat detector technology will to some degree influence, and perhaps replace, current imaging systems that combine an image intensifier and spotfilm device for cassette images. Required dose and image quality, however, will be the primary drivers of change.

Multifunctional C-arm fluoroscopy systems

Fluoroscopy-guided exams from numerous projections using either an undertable or overtable imaging system with simultaneous table tilt (Fig. 6.17-2 and 3) can be performed using a C-arm to support the imaging system.

6 X-ray Systems for Diagnostics and Intervention

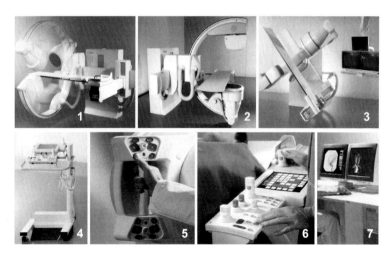

Figure 6.17 Multifunctional C-arm X-ray system for universal examinations in internal medicine, vascular diagnostics and interventions

A C-arm system is also able to acquire "rotation images" (Fig. 6.17-1) for calculating 3D images (see section 9.4). A remote-controlled fluoroscopy system with image workstation and network connection (7) can also be operated close to the patient (5). For interventional exams, the units that control system movements and image parameters can be positioned directly beside the patient table (6). A mobile operating unit (4) can also be used.

6.3.7 X-ray systems for urology

Under certain conditions, general urologic exams can be performed using universal fluoroscopy systems or X-ray image systems, if the exam room has been appropriately prepared and equipped.

Special and comprehensive exams of the urogenital system, however, do require X-ray systems specifically designed for that purpose. For urologic exams of the kidneys, renal pelvis and urinary bladder or follow-up exams, sonographic imaging systems can also be used, however a plain-film radiograph of the abdomen is always created as the primary X-ray image.

System variations

To delineate the kidneys and urinary tract, a contrast agent is administered intravenously for an excretory urogram. Retrograde cystography is a radiological examination of the bladder in which a contrast agent is inserted in the bladder via a catheter (see section 9.3 for contrast media).

6.3 Technology of X-ray Systems

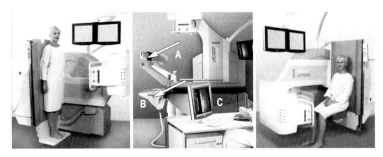

Figure 6.18 X-ray system specifically designed for standard and special urologic diagnosis and treatment, endourologic exams and minimally invasive surgical interventions. Imaging workstation with network connection (C).

Patient positioning requirements for the urography X-ray system are unique. The patient sits comfortably on an anatomically designed seat located at the end of the table (Fig. 6.18, arrow A). A urinary receptacle can be attached to the end of the patient table, if needed (Fig. 6.18, arrow B).

The imaging system must be able to produce an overview image of the abdomen showing the renal pelvis, renal calices, ureters, bladder and genital area. For excretory urography, the patient table is placed in the vertical position, allowing the patient to be diagnosed while seated comfortably on the micturition seat (Fig. 6.18, right).

The table has to be extendible for exams and interventions requiring the patient to lie prone (Fig. 6.19, arrow).

X-ray systems for urology can capture images directly and are usually equipped for dynamic imaging. For dynamic studies, an image intensifier with an X-ray television screen is attached to the spotfilm device. As solid-

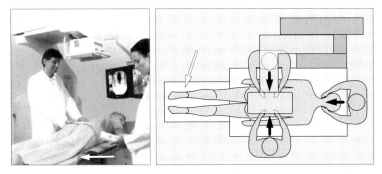

Figure 6.19 The patient has to be easily accessed from all sides for minimally invasive surgical interventions

state detector technology matures, however, it will continue to replace the image intensifier in urological imaging systems.

The image receptor system or spotfilm device contains an image intensifier, an assembly for cassette images (film-screen and/or imaging plate), a device for collimating the image format close to the cassette, a grid for reducing the scatter radiation, and a measurement chamber for automatic exposure.

The degree to which flat detectors alter and replace current systems based on a combination of image intensifier and spotfilm device for cassette images depends on how the technology develops but, more importantly, on required dose and image quality.

Modular system variants

One alternative X-ray system for the urology practice or hospital department is a configuration of modular subsystems (Fig. 6.20). The largely mobile system components are advantageous because each system component can be targeted for specific diagnostic and therapeutic requirements, e.g., in general urological diagnostics, endourology, lithotripsy, prostate diagnosis and treatment in the operating room, for bedside images, or angiographic exams.

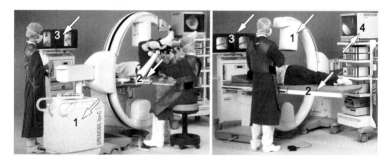

Figure 6.20 Modular diagnosis and treatment system

In a modular design (Fig. 6.20), the system components consist of a mobile C-arm fluoroscopy system (1); a mobile patient table (2) for urological and other exams, such as minimally invasive interventions or treatment of stones; a mobile image-viewing and analysis workstation (3); a cart with systems for all endoscopic exams, including a monitor (4) and a mobile ESWL system (Fig. 6.21, right).

X-ray systems for ESWL

Lithotripsy is the disintegration of kidney, bladder or gall stones. Kidney stones used to be removed surgically by means of a cystoscope inserted through the urethra. This procedure is still used occasionally. The primary

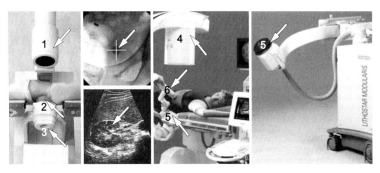

Figure 6.21 X-ray imaging and treatment system suitable for ESWL and endourologic exams (left) and a modular system with lithotripsy functionality

procedure, however, is extracorporeal shock wave lithotripsy (ESWL), a virtually pain free and gentle procedure that disintegrates stones.

X-ray fluoroscopy systems designed to pinpoint stones were developed especially for ESWL. A C-arm imaging system with an image intensifier precisely localizes the stone using two angle positions around the isocenter. The shock-wave head is moved into position by lining it up with the central X-ray beam (2). The position of the stone can be continuously monitored using an ultrasound probe (3).

As an alternative, the modular system discussed above can also be used for examination, stone localization and treatment (Fig. 6.21, right).

Modular systems also use an image intensifier on a mobile C-arm to localize stones. A mobile lithotripsy unit (right, with shock-wave head) is linked to a C-arm and is positioned using the shock-wave head. An image intensifier and/or ultrasound system is used for localization and positioning (6).

6.3.8 X-ray systems for special angiography

Angiography is the radiological imaging of the arteries (arteriography) and venous system (phlebography) after injection with a contrast agent. Angiography differentiates between selective and non-selective procedures.

- Non-selective or surveillance angiography:
  The contrast agent is injected into the main artery (aorta abdominalis or aorta thoracica). Large vessels and their small afferent arteries are shown.
- Selective angiography:
  An image of the artery and its organ system is generated.
- Superselective angiography:
  The small and smallest arterial branches are viewed, e.g., during interventions.

6 X-ray Systems for Diagnostics and Intervention

Cardiac angiography

Cardiac angiography or angiocardiography is the radiological imaging of the interior of the heart, the large neighboring arteries and the coronary artery.

Neuroradiology

Neuroradiology is the radiography of the skull, including the brain, the arteries and veins of the neck, and the spinal canal after injection with contrast agent.

System variations

All modern radiography systems for general and special angiographic, cardiac angiographic and neurological examinations are designed as C-arm systems.

The image acquisition system with the image receptor (here: image intensifier) and collimator is attached to the C-arm at a tangent to the X-ray tube so that the central ray passes through the geometric center of rotation (isocenter) of the imaginary full circle (Fig. 6.22, left).

The C-arm system has been the imaging system of choice for specialized angiography for many years. C-arm systems offer major diagnostic advantages:

- The region under study is always located in the isocenter (ISO) of the imaging system, regardless of the many different projections allowed by the gimbal-mounted C-arm.
- Image projections that are easy to set and modify also enable fast imaging without superimposition.
- Orbital and angular movements of the C-arm permit an image mode that calculates a 3-D image using a special software program.

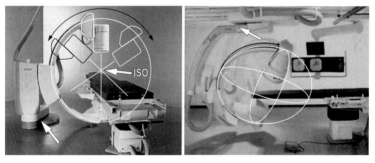

Figure 6.22 X-ray system for universal angiographic applications. Floor-mounted (left) and ceiling-mounted (right) single-plane C-arm system with image intensifier.

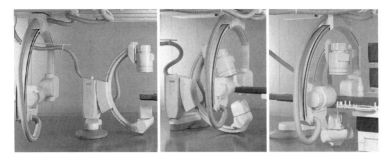

Figure 6.23 Biplane C-arm system. Left: Ceiling-mounted C-arm in park position, e.g., for patient preparation or single-plane examination. Center: Position for cerebral vascular system and spinal canal exams or for interventions. Right: Position for peripheral or abdominal exams or interventions.

Universal angiography primarily uses single-plane angiography systems (monoplane), i.e., a C-arm with an image acquisition system. For special exams, such as in cardiology or neuroradiology, biplane angiography systems are mainly used, e.g., to enable direct viewing during interventions of projection planes shifted by approximately 90°. X-ray angiography systems for universal applications must allow the patient to be examined from head to toe without repositioning.

As circumstances and diagnosis require, the C-arm of the single-plane angiography system (monoplane) can be mounted on the floor or the ceiling (Fig. 6.22, arrow). The major advantage of the ceiling-mounted C-arm is its virtually unrestricted patient access. In biplane angiography systems, one C-arm is mounted on the floor, the other on the ceiling (Fig. 6.23).

Magnetic navigation

In electrophysiology and interventional cardiography, the physician inserts the catheter by hand and guides it to the site of interest. Because the heart is a complicated organ, automatic navigation provides a way to optimally guide the catheter to the site of diagnosis or treatment without putting the patient at risk.

During fluoroscopy, the position of the cardiac catheter is recorded at all times. The physician inserts the catheter by hand and a computer program, using the magnet systems positioned on the patient's left and right, helps to guide the tip of the catheter to the point of interest (Fig 6.24-2).

For this type of intervention, it is essential that the patient, physician, and the proximate system components are absolutely tolerant of magnetic fields. For this reason, X-ray systems that use magnetic navigation may not be used for patients with pacemakers or metallic implants; nor are they suitable for use

6 X-ray Systems for Diagnostics and Intervention

Figure 6.24 Single-plane C-arm system with dynamic flat detector (1) and two magnetic devices for magnetic navigation of the cardiac catheter (2)

with magnetically sensitive components such as image intensifier tubes and CRT monitors.

Measuring stations

Radiological exams of the heart and coronary arteries are performed in specially equipped cardiac catheter work rooms, known as cardiac catheter labs (Fig 6.25).

Because the patient being examined is almost always suffering from pathological changes to the condition of the heart and its vessels, the equipment must be highly sensitive. Extremely delicate measuring devices, such as hemodynamic and electrophysiologic recording and information systems, support the examination team (Fig. 6.26).

Automatic computer analysis is used to detect and calculate the contours of coronary arteries or create comprehensive analyses to assess the functionality of the left ventricle.

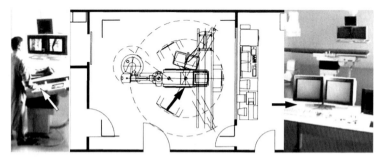

Figure 6.25 Left: example of a single-plane cardiac cath lab (exam room) with tableside operation of all the required system and imaging parameters. Right: image processing and reporting workstation with network connection (control room).

6.3 Technology of X-ray Systems

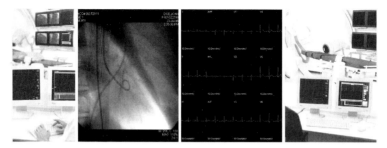

Figure 6.26 Recording and information system for performing all hemodynamic and electrophysiologic exams in the cardiac cath lab

Electrophysiologic instruments are used to measure and map cardiac action potential. Highly sensitive, intracardiac electrode catheters are inserted through the veins into the right ventricle.

Hemodynamics includes the physiological conditions or pathological changes of the circulation of blood through the vessels, such as abnormal blood pressure, volume, flow mechanics, vessel elasticity, etc.

All of today's digital imaging systems can be connected to a network, as can all the systems previously discussed. Cardiac cath labs usually have their own bus network that is connected to the PACS and the HIS (see chapter 10).

6.3.9 X-ray systems for surgery

Surgery is the branch of medical science that treats diseases and conditions in the organs, the skeleton and the circulatory system through surgical interventions using operative or manual procedures. Surgery is divided into a number of specializations:

- Thoracic and visceral surgery (viszeralis, related to the viscera)
- Cardiac and vascular surgery
- Emergency surgery (traumatology)
- Neurosurgery (neurology: diagnosis and treatment of diseases of the nervous system and musculature)
- Pediatric surgery
- Oral surgery
- Plastic surgery (reconstructive or cosmetic surgery)

To provide support in the operating room, the C-arm system has to cover a broad range of radiologic disciplines. It must be able to visualize minimal contrast differences, e.g., for kidney and gall stones, long shadows, etc., to detect bony structures and fractures or artificial limbs, and last but not least, visualize blood vessels or ventricles during cardiac operations.

System variations

The mobile C-arm system or surgical image intensifier has proven to be an effective tool in the operating room (Fig. 6.27). These systems generally have two monitors. The first monitor always displays the current fluoroscopy image, or live image. The second monitor, known as the reference image monitor, displays the images that were saved during fluoroscopy (reference images).

"Last Image Hold (LIH)" is a feature that retains the last live image acquired before the fluoroscopy is turned off and displays it on the monitor. All X-ray systems with digital imaging offer these functions (see chapter 9).

The first X-ray systems with C-arms were developed in the 1950s, when the mobile image intensifier began to be used in operating rooms for fluoroscopy during surgery.

Like the mobile X-ray systems for bedside imaging (see section 6.3.3), the surgical C-arm is a fully independent X-ray unit. It has its own generator and all the components needed for fluoroscopy. The image intensifier and the X-ray tube are mounted opposite one another on the C-arm. The X-ray tube assembly is also referred to as a single tank, i.e., the tube and generator form a single unit (see chapter 7).

A cassette holder can be placed over the input screen of the image intensifier, enabling the acquisition of targeted images. It is only a matter of time, however, until the flat detector supplants this image intensifier in surgery as well.

In addition to the mobile and positionable surgical C-arm fluoroscopy system, the ceiling-mounted C-arm can be used in the operating room. Unlike the ceiling-mounted, permanently installed C-arm, however, the mobile C-arm system can be used in other operating rooms, intensive care and emergency.

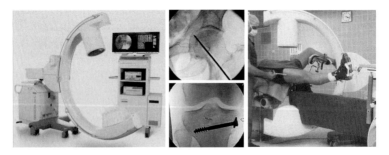

Figure 6.27 Mobile C-arm or surgical image intensifier

6.3 Technology of X-ray Systems

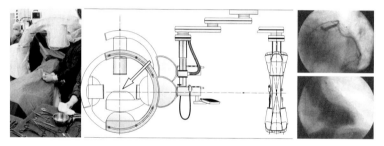

Figure 6.28 Ceiling-mounted surgical C-arm

There are C-arm systems with and without an isocenter (isocenter Fig. 6.28, arrow; see C-arm systems for angiography and chapter 9). A C-arm system with an isocenter is particularly beneficial in cardiologic exams. The area of interest remains in the same location on the monitor, regardless of the angle settings of the X-ray central beam. It is also possible to calculate 3-D images using angle-triggered image modes.

Mobile C-arm systems without an isocenter, on the other hand, are generally lightweight and easier to move. Because the construction of the system is not tied to an isocenter, the penetration depth is greater, i.e., the distance of the central ray to the inner edge of the C-arm is larger.

Of course, surgical C-arm systems with digital imaging can also be connected to a network. Like cardiac cath labs, C-arm systems usually have their own bus network that is connected to the PACS and the HIS (see chapter 10).

6.3.10 X-ray systems for computed tomography

Computed tomography (CT) is a computer-supported slice imaging procedure for X-ray diagnostics.

In the 1960s, the English engineer Godfrey N. Hounsfield (1979 Nobel prize) developed a new imaging procedure for generating non-superimposed transverse slice images by rotating the image acquisition system 360° around the longitudinal axis of the patient. Hounsfield is the inventor of computed tomography, which is considered the greatest innovation in the field of radiology since the discovery of X-rays in 1895 by W.C. Röntgen. The first clinical CT images were captured in 1972 in London at Atkinson Morely's Hospital.

The major advantage CT has over conventional imaging techniques is the ability to acquire consistently non-superimposed images by rotating the image acquisition system. In this way, the many different projections of the same slice produce a distortion-free and higher-contrast image of all the volume elements (voxel) located in this transverse slice.

6 X-ray Systems for Diagnostics and Intervention

Figure 6.29 CT exam and control room with image analysis and image processing station with network connection. CT exam (right).

CT is used to support sonographic, nuclear medicine and conventional X-ray diagnostics (including magnetic resonance imaging), especially to detect circumscribed and diffuse morphological changes, e.g., in tumors, metastases, abscesses or diseases of the lymphatic system.

CT supports, e.g.:

- The detection of strokes, head injuries, intervertebral disk prolapse, abscesses
- The localization of fractures
- The determination of the severity of skeletal and soft-tissue injuries in trauma patients
- The diagnosis of pathological changes in various organs
- Diagnosis through exclusion of other diseases

This book covers only the fundamental differences between CT and conventional X-ray imaging technology.

- For a more complete discussion of computed tomography, see W. Kalender in section 11.3 under References.

Like all X-ray systems, the image acquisition system has an X-ray tube and an image receptor system. These are attached tangentially opposite to one another and housed in a ring-shaped examination device, the gantry. The X-ray central beam passes through the center (the isocenter) of the geometric circular arc rotating around the patient (Fig. 6.30, left).

Detector and collimation

The image receptor system consists of a number of detector elements that are arranged around a circular opening, the detector ring. This detector system converts the impinging X-rays of differing intensity into electrical signals (see section 8.5, flat detection systems). Downstream electronic components intensify and digitize these signals (see section 9.2). By rotating the imaging

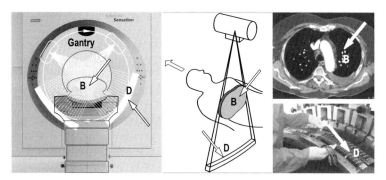

Figure 6.30 CT unit with imaging system in a ring-shaped acquisition unit, the gantry. CT detector assembly (bottom right).

system around the patient, a large number of different projections are generated. The resulting projection data are fed into a computer that calculates and generates the image (transverse slice B).

The X-rays emitted by the tube are appropriately collimated to the maximum required beam cone, as in all X-ray systems. Similar to X-ray acquisition systems with collimation close to the cassette or detector, the cone beam is collimated close to the detector in CT. This reduces the scatter radiation on the detector system and related image artifacts.

The detector system consists of multiple, adjacent detector rows known as multiline detectors. They are narrower in the middle (as related to the longitudinal axis of the body) and become wider further out. They are called adaptive array detectors. The combination of collimation and electronic shuttering of the detector rows provides a great deal of flexibility in selecting slice thicknesses (Fig 6.30-D).

Sequential CT

In sequential CT, a transverse slice is imaged each time (Fig 6.30-B). The imaging system rotates 360° in a process known as a scan. In order to scan the region of interest completely, several adjacent transverse slices are acquired. After acquiring the initial slice, the patient is moved a defined distance lengthwise (usually the slice thickness), and the next scan is performed.

If the patient moves during the acquisition, the data from the various angle positions will not align for a precise reconstruction of the image. The image exhibits motion artifacts and therefore has only limited diagnostic value. For this reason, the slice imaging procedure of sequential CT is not particularly useful for diagnosing anatomical regions subject to periodic movement, e.g. the heart and lungs.

Spiral CT

The introduction of slip-ring technology in 1987/88 made possible the continuous rotation of the imaging system. This represented a clear break from conventional CT and the slice acquisition procedures it used.

Unlike sequential CT, the patient is continuously moved through the image acquisition system along the longitudinal axis (Fig. 6.31-Z). The image acquisition system (X-ray tube and detector) performs a number of 360° rotations (scans) always in the same direction of rotation. The area of interest is scanned in the shape of a spiral. The latest CT systems (spiral CT) need only 0.4 seconds to complete one rotation of 360°.

By continually moving the patient longitudinally, a virtually spiral-shaped scan of slices is acquired to create the image. The image acquisition system (tube and detector) always rotates geometrically around a point on the z axis.

CT uses primarily the transverse plane (Fig. 6.31-T) as the image plane (see section 9.4 for more information on the transverse plane). For this reason, views with a different orientation usually have to be calculated from the original images. The procedure used for this purpose is called multiplane reformatting (MPR).

A series of axial images are "stacked" together. By lining up the same image row in the series, any plane can be calculated from the adjacent images. So transverse plane (Fig. 6.31-T) views can be used to create axial, sagittal, (Fig. 6.31-S), coronary (Fig 6.31-C) or oblique sections.

A special computer program can also calculate and display 3-D images from the available data. Using the data volume from the CT values, a spatial image can be generated voxel for voxel (volume element). These virtual views are especially good for structures that stand out from their surroundings, such as the skeleton or vessels filled with contrast agent.

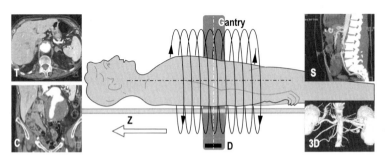

Figure 6.31 The principle of spiral CT

CT angiography (CTA)

The extremely short scan times permitted by slip-ring technology brought about the multislice scanner. A multislice scanner is able to display the entire vascular system down to the tiniest structures at maximum contrast agent enhancement.

Image post-processing allows for visualization of the entire vascular system, even tiny vascular branches and emboli. Data volumes can also be used afterwards to create 3-D images from any beam direction, e.g., for surgical planning.

7 X-ray System Components

Every X-ray system, regardless of the radiological diagnostic discipline, is comprised of the following system components:

- Basic system for positioning patients and setting exposure projections (Fig. 7.1-1, section 7.1)
- X-ray generator for generating high voltage (Fig. 7.1-2, section 7.2)
- X-ray tube for generating X-rays (Fig. 7.1-3, section 7.3)
- Primary collimator for focusing and collimating the X-rays (Fig. 7.1-4, section 7.4)
- Image receptor system (Fig. 7.1-5 and chapter 8): Film-screen/imaging plate, image intensifier, flat detector
- Floor or ceiling-mounted stand (optionally for a second exposure plane) (Fig. 7.1-6, section 7.6)
- Image processing and evaluation station (Fig. 7.1-8 and chapter 10)
- Image viewing station (Fig. 7.1-7, section 8.8):

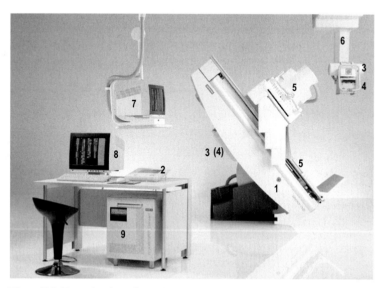

Figure 7.1 Example of a radiography system (undertable fluoroscopy system)

- Archiving unit, analog and/or digital
- Communication system with internal or external network connection in the case of a digital X-ray system (Fig. 7.1-9 and chapter 10)

Regardless of the type of imaging system and the components used in creating an effective diagnostic visualization, every radiological exposure must be processed via the image receptor system. Only then can the result be displayed, communicated, stored and archived.

Today's digital world provides almost unlimited possibilities for communicating diagnostic information. Numerous solutions are available – from internal hospital networks to worldwide communication via the internet (see chapter 10).

7.1 Basic System

To produce effective X-ray images, the object to be recorded (i.e., the patient) must be positioned correctly with regard to anatomical conditions and the necessary exposure projection(s). The incident radiation beam must project through the region of interest and onto the image receptor system.

Specific radiography systems are used and the patient is positioned based on radiological requirements. X-ray systems are considered a unit comprised of the basic system, generator, tube and image acquisition system. Exceptions to this include non-floor-mounted systems, such as the mobile surgical C-arm which is moved to the examination or operating table and positioned there. This is similar with mobile radiography systems for bed exposures but varies with systems for trauma or emergency exposures.

7.2 X-ray Generator

The X-ray generator is the "heart" of a diagnostic X-ray system. It converts the current supplied by the network to the required electrical energy as adjustable tube and exposure voltage U_R and the corresponding high-voltage waveform for generating X-rays (see section 4.2).

The X-ray generator

- Regulates the heat current in the cathode for electron emission (Fig. 7.2-H)
- Generates the necessary high voltage for accelerating the electrons between the cathode and the anode

7 X-ray System Components

- Supplies the three-phase current for driving the rotating anode (Fig. 7.2-N)
- Makes the exposure using external dose measuring devices (I), e.g., automatic exposure controls for film-screen or imaging plate systems or automatic dose rate control during fluoroscopy (Fig. 7.2-Q)
- Calculates the data for protecting the X-ray tube from overloading and
- Stores the data protocol, such as exposure data, error diagnosis, etc., in the central processing unit (Fig. 7.2-CPU, central processing unit).

The mechanical, electrical and electronic components of an X-ray generator are accommodated in a control cabinet. Only the user interface of the generator is still located in the examination or control room (Fig. 7.2-G). The generator and system controls form one unit in the case of most radiography systems.

7.2.1 Multipulse generators

Even if the previous generation of conventional 2 through 12-pulse generators are still produced and used around the world, only multipulse generators are used in new installations for modern radiodiagnostics in developed countries.

Multipulse generators function according to the principle of high-frequency inverter technology, also known as direct voltage conversion (Fig. 7.2).

The necessary power connection (N) is typically: 400 V ±10% three-phase current, 50/60 Hz ±1 Hz, or 440/480 V ±10% three-phase current, 50/60 Hz ±1 Hz via an internal pretransformer.

Figure 7.2 shows the principle course for generating high-frequency tube voltage U_R: The initially rectified (A) and smoothed (B) alternating voltage is converted using a direct current rectifier (C) into a high-frequency

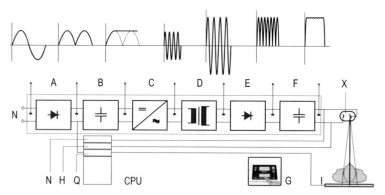

Figure 7.2 Principle of multipulse or high-frequency generators

7.2 X-ray Generator

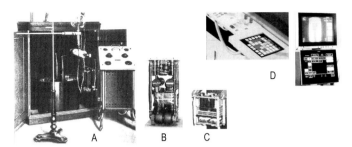

Figure 7.3 First three-phase current X-ray generator from Siemens in 1915 with a generator control panel (A); conventional 12-pulse generator (B), and 100-kW RF generator (C) with modern touchscreen control panels (D) integrated in the system controls or in the image processing station. A, B, and C show approximately real-size relationships.

alternating voltage (D). This is then rectified (E) and smoothed (F) again and is available as high-frequency multipulse tube voltage U_R for generating X-rays in the X-ray tube (X).

Multipulse generators, also called high-frequency/RF generators or converter generators, meet the requirements for minimal, negligible waviness and thus for particularly brief switching times (up to approx. 1 ms) and optimal radiation hygiene.

In contrast to conventional generators, RF generator technology allows a significant reduction in weight and size, thereby allowing a smaller, more compact design (Fig. 7.3).

X-ray generators and tubes can be combined in a common housing for diagnostic applications using less power. These are known as self-contained or single-tank X-ray generators (Fig. 7.4, arrow).

Figure 7.4 Conventional 2-pulse generator: transformer T with X-ray tube X (left). Self-contained or single-tank RF generator: Transformer with tube in one housing and self-contained RF X-ray generator with X-ray tube and collimator on an exposure stand (arrows).

107

All required types of digital imaging and data control can be realized with microprocessor-controlled RF generators. Information distribution is controlled via a central computer equipped with microprocessors using a digital code (CPU).

Mobile generators

Mobile X-ray systems, such as mobile generators for bed exposures, are connected to conventional power sockets for electrical voltage supply. Since these sockets cannot provide electrical energy sufficient for the necessary radiographic output, it must first be stored in a capacitor. This function is comparable to that of the capacitor in a flash photo device: Charging prior to exposure and discharge, i.e., transfer of the stored total energy in fractions of a second, during exposure. For additional information on mobile generators, see section 6.3.

The advantage of energy storage is that the capacitor is situated on the primary side of the high-voltage circuit. When the capacitor discharges during the exposure, it is possible as a result to adjust the tube voltage and keep it constant. This represents a radiographic advantage over configuration on the secondary side of the high-voltage circuit, since the tube voltage consequently decreases sharply during the exposure. For additional information on exposure data and automatic exposure controls, see chapter 9.

7.3 X-ray Tube

The X-ray tube (Fig. 7.5) is the "extended arm" of the X-ray generator. The tube voltage U_R generated in the X-ray generator is supplied via the high-voltage cables to the X-ray tube to generate the X-rays.

Figure 7.5 shows: The X-ray tube and protective housing form one unit and are referred to as the X-ray tube or the tube (X). The primary diaphragm or multileaf collimator (P) is directly connected to the X-ray tube. It is required for collimating or sizing the useful beam (see section 7.4). Low-energy radiation is pre-filtered in the collimator (see chapter 4).

Generator and tube power data must be consistently compared according to the X-ray power necessary for radiological imaging. The technical and physical features relevant for radiography, such as focal spot power and size, anode angle and diameter, or tube design, etc., differ according to the particular diagnostic use required (see section 7.3.1 through 7.3.3).

Figure 7.6 shows the basic principles of an X-ray tube: Tube with rotating anode (3), Wehnelt cylinder (13) as the cathode and protective house (6). The radiation emitting from the focal spot (focus) is depicted here in a general

7.3 X-ray Tube

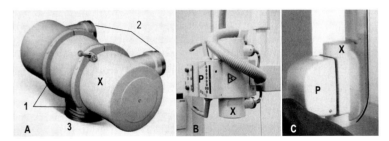

Figure 7.5 A: X-ray tube (X), assembly rings (1), plug heads for the cathode and anode-side high-voltage cable, as well as connections for the heat current, and stator (2) and X-ray window with a flange for attaching the collimator (3). B: X-ray tube (X) with collimator (P) on a ceiling-mounted tube stand. C: X-ray tube (X) with collimator (P) on an angiography C-arm system.

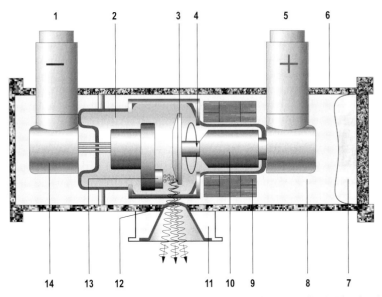

Figure 7.6 Schematic design of a standard rotating anode X-ray tube: 1) Plug head for cathode-side high-voltage and heat current cable; 2) glass envelope of the rotating anode tube; 3) rotating anode; 4) metal ring instead of glass for glass-metal center part tubes; 5) plug head for anode-side high-voltage cable and stator cable; 6) protective tube housing; 7) membrane for oil expansion; 8) coolant (e.g., oil); 9) stator for anode drive; 10) rotor; 11) flange for attaching the collimator; 12) X-ray window (e.g., beryllium, atomic number 4); 13) cathode (Wehnelt cylinder); 14) tube supports.

manner as if directed in a targeted manner through the X-ray window (12). The X-ray quanta are actually emitted to all sides from the focal spot. The protective tube housing (Fig. 7.5-X and 7.6-6) has the following functions:

- Holding and securing the tube to the tube stand
- Attaching the collimator (11)
- Accommodating the cable feeds for high voltage (1 and 5), heat current, and rotor drive, etc.
- Enclosing in lead for radiation protection (minimizes leakage radiation, housing leakage radiation)
- Insulating against high-voltage arcing

For improved heat dissipation (overload protection), the housing is filled with oil (8), for example, and is sealed at all line feeds (high voltage, heat voltage, stator voltage) and at the radiolucent exit window such that no fluid can escape. This sealing also provides protection against possible implosion (vacuum loss).

An X-ray window (12) limits the desired useful beam and filters low-energy radiation at this location.

7.3.1 X-ray tube

The X-ray tube is a virtually air void tank made of glass, glass-metal/ceramics or all metal (see Straton™ tube) in which X-rays are generated by adjusting the cathode and anode (see chapter 2). The vacuum of the X-ray tube is a high vacuum in which an air pressure of 10^{-6} to 10^{-7} mbar (millibar) prevails.

X-ray tubes have different designs and output classes, according to the radiological requirements. They are designed for all radiography systems used in modern radiology – from endoral exposures in dental radiography to high-performance tubes for cardiac angiography and CT systems (Fig. 7.7).

In the principle rotating anode X-ray tube shown in Figure 7.6, the rotor of the rotating anode is supported by ball bearings. However, ball bearings are subject to mechanical wear. To protect the ball bearings and to achieve the longest possible service life, the rotating anode is accelerated to a rotational speed of more than 8,000 revolutions per minute or rpm (not during fluoroscopy) prior to every exposure via a starter in the case of a rotating anode X-ray tube.

Design engineers sought to create a rotating anode X-ray tube with a rotor without ball bearings. The solution included anode rotation, in which the rotor turns on liquid-metal liquid bearings.

7.3 X-ray Tube

Figure 7.7 X-ray tubes (not shown to scale): high-performance tube with (1) heated rotating anode and (2) glass-metal design; (3) rotating anode X-ray tubes for mammography; (4) for all diagnostic routine examinations using conventional or digital radiography systems; (5) stationary anode tube, e.g., for endoral exposure techniques (dental) and (6) first rotating anode tube from 1933 for fluoroscopic and radiographic operation.

Liquid-bearing technology

The principle of the liquid-metal liquid bearing (also known as a spiral flute bearing or spiral liquid bearing) can be explained on the basis of the vehicle aquaplaning effect on wet surfaces.

In 1964, E.A. Muijderman introduced the spiral liquid bearing in his dissertation at the Technical University of Delft, Holland. In 1989, the first tube supported in such a manner was used by radiographic systems in cardiac angiography.

Figure 7.8 shows the copper rotor (5) which is permanently attached to the hatched support parts (3) for the anode (2). The rotor is driven by the rotating magnetic field generated in the stator (6). A particularly small intermediate space of 15-20 μm filled with a liquid metal is located between the rotating and the stationary support parts. When the rotor is driven, this liquid metal is distributed and forms a liquid wedge (arrows) thereby causing the rotating support part to "hydroplane" on the stationary bearing shaft (5) of the rotor.

The stationary bearing shaft is permanently attached to the tube housing (1). The interior is hollow so that coolant (7) can flow directly through thereby effectively dissipating the heat.

In contrast to the rotor bearings using ball bearings, the rotors (rotating anode tubes) using liquid-bearing technology function in an extremely low-

7 X-ray System Components

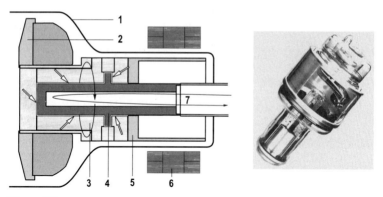

Figure 7.8
Cross-section of the rotor with the rotating anode using liquid-bearing technology

noise manner, which makes work more pleasant for all those involved; this is particularly beneficial during longer examinations such as during interventions using angiography systems. Liquid bearings function with almost no wear as opposed to ball bearings. Therefore, the rotating anode can rotate at consistent high speeds with almost no limitation of the service life and does not need to be accelerated to the required rotational speed prior to triggering radiation.

Straton™

In the first several years following the discovery of X-rays, it was customary to decelerate the X-ray quanta emitted by the cathode at the glass envelope of the X-ray tube. Particularly high heating occurred at this location, i.e., the focal spot. To achieve improved heat distribution, American R.W. Wood proposed as early as 1897 rotating the glass tube receptacle about its center axis (Fig. 7.9, left). This was to achieve improved heat dissipation via cool air about the entire glass tube body, i.e., the heat would be dissipated directly at the virtually enlarged point of origin. This idea was initially insignificant. It was not only extremely complicated to implement this idea due to the technological possibilities of the time but the predecessor of the modern rotating anode was also already being developed.

At the end of the 19th century, in the first years of practical radiography, tube design was still in the beginning stages. The glass envelope of the tube was also the "anticathode." It was responsible for collecting and decelerating the electrons and became very hot as a result. The brittle and poorly heat-conducting glass was not favorable for such thermal loads. Consequently, the performance and service life of the first X-ray tubes were unsatisfactory.

7.3 X-ray Tube

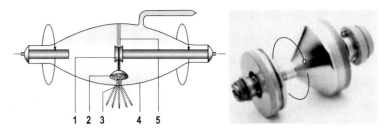

Figure 7.9 Rotating anode X-ray tube with glass housing 1897 (left) and metal housing 2003 (right)

Figure 7.9 shows the basic design of an X-ray tube in 1897: the cathode (2) swivel-mounted on the bearing cone (1) hangs down under its own weight. The electrons strike the focal spot (3) of the rotating glass envelope, which was also the anticathode (4), and form a focal spot path around it (5). During rotation, the heat from this focal spot path spreads over the entire glass body and can be dissipated in a significantly more effective manner.

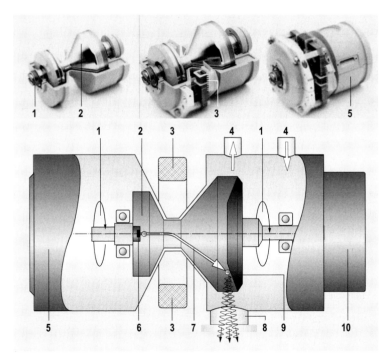

Figure 7.10 Design of the Straton™ rotating anode X-ray tube

113

However, the dissipation of heat continues to be a constant challenge in the development of X-ray tubes. After over 100 years, the design engineers of Siemens Medical Solutions implemented this idea and produced the first X-ray tube based on the R.W. Wood's idea.

Figure 7.10 shows the basic design of the Straton™ X-ray tube with a directly cooled anode. The metal X-ray housing (2), the anode (9), and the cathode with the heater plug filament (6) form one unit, which is supported by a shaft (1) and held at a constant rotation speed by a motor (10).

Figure 7.10 shows the electrons emitted by the cathode accelerated in the direction of the anode in voltage gradient U_R. To bring the electron current to the intended focal spot (9) and to keep it there during the constant rotation, specially adjusted coils (3) permanently direct the electron current (7) to the location of the anode provided for the focal spot (focus). The X-ray quanta resulting at the anode leave the X-ray tube (5) via the X-ray window and the flange for attaching the collimator (8). The coolant in the X-ray tube housing (5) is continuously replaced (4). It surrounds the tube unit and cools the anode directly.

7.3.2 Cathode

The cathode is the source of the electrons. A coiled tungsten wire with a diameter of 0.2 to 0.3 mm, also known as a spiral-wound filament (a heater plug element in the case of the Straton™ CT tube), serves as the electron source and heats up to between 2,000°C to 2,600°C via an electrical current (I_H = heat current).

The tungsten wire also forms the tube cathode. The significant heating of the tungsten wire releases and emits a number of electrons (visually comparable with the processes in a light bulb). The higher the temperature of the spiral-wound filament, the more electrons emitted. They form tube current I_R. The heating output therefore determines the size of tube current I_R and consequently the quantity (dose rate) of the radiation.

Figure 7.11 shows the configuration of the spiral-wound filament and the heater plug filament. The cathode can be made of one, two or three spiral-wound filaments (or one heater plug filament). They are embedded in a focusing cup (5). Depending on how the focus and consequently the useful beam are to be generated for the diagnostic application, the spiral-wound filament can be configured differently (Fig. 7.11, 1 to 4).

To direct the electron flow to the desired focal spot of the anode, it must be focused. This is achieved by a cylindrical focusing cup (5), the Wehnelt cylinder (see appendix).

The cylindrical form creates an electron space-charge cloud. The electrons are accelerated in the potential gradient (U_R) of the tube from the Wehnelt

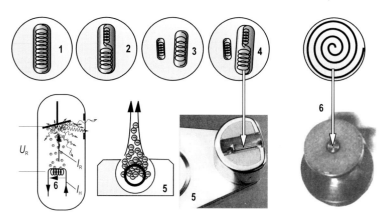

Figure 7.11 Schematic diagram of electron focusing using a Wehnelt cylinder and a heater plug filament (right)

cylinder toward the anode in a targeted manner (tube current I_R; see section 2.1, Fig. 2.2, the generation of X-rays in a vacuum tube).

Without the Wehnelt electrode, the electrons would impinge on the anode over a large area, and would not be concentrated in one focal spot. The effect of the Wehnelt electrode depends on its potential (its electrical voltage) with respect to the spiral-wound filament (heater plug filament), i.e., the incandescent filament.

Grid-controlled X-ray tube

Radiographic systems with high image frequencies, as required in angiography and, especially, in cardiac angiography, must perform a large number of exposures with very short switching times (and switching intervals); this is referred to as a pulsed exposure technique (see section 9.2).

To switch the radiation in pulsed electron emission periods for such image series, the Wehnelt cylinder is equipped with a grid control and is used as a control electrode for pulsed radiation, i.e., radiation pulses switched at certain time intervals (Fig. 7.12).

a) The Wehnelt electrode and spiral-wound filament have the same voltage potential, thereby focusing the electrons on the focal spot: implemented in all standard tubes.

b) The Wehnelt electrode is negative with respect to the spiral-wound filament. The electrons are focused on an even smaller focal spot. Application: generation of a smaller third focus size for some 3-focus tubes (e.g., in cardiac angiography or neuroradiology).

7 X-ray System Components

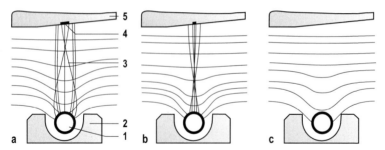

Figure 7.12 Basic diagram of electron focusing with grid control: cathode, incandescent filament (1); Wehnelt electrode (2); electron beam (3); focal spot, focus (4); anode segment (5).

c) The Wehnelt electrode has a higher negative potential (typically approx. - 2 kV) with respect to the spiral-wound filament. Despite the applied tube voltage, no electrons can reach the anode. The tube is "blocked" and has no space-charge effect. By changing the voltage at the Wehnelt cylinder, the tube current I_R is switched on and off in a completely delay-free manner. Application: "grid-controlled" tubes.

The Wehnelt cylinder which is normally at the cathode potential is switched in this process at the switching intervals to the anode potential. The electron emission is interrupted in this time. There is no space-charge effect. The advantage is virtually inertia-free control of the radiation pulses. This is referred to as a secondary pulsed exposure technique and the tubes are referred to as grid-controlled X-ray tubes.

An alternative is the primary or generator-side switched pulsed exposure technique. High voltage: On/off. It is used mainly for pulsed exposure techniques.

Space-charge effect

To reduce the space-charge cloud in the X-ray tube, i.e., to move (accelerate) the emitted electrons toward the anode, a certain voltage potential is necessary (between the cathode and anode). When the anode potential is reduced with respect to the cathode or tends toward 0, the space-charge effect of the anode potential on the cathode is reduced or blocked.

7.3.3 Anode

The anode of the X-ray tube can be stationary (stationary anode = stationary anode tube) or a rotating disk (rotating anode = rotating anode tube).

Keep in mind that the electrons emitted from the heated tungsten wire, the spiral-wound filament, are accelerated toward the anode by the voltage applied between the cathode and anode (U_R = tube voltage, also potential or potential gradient) (e.g., at U_R = 100 kV) to approx. 165,000 km/s. The point of impingement is referred to as the (electrical) focal spot or focus (Fig. 7.13-7 and 7.14).

Stationary anode tubes

Stationary anodes have a smaller design and can only be loaded to approximately 2 kW. They are used, for example, in mobile generators or for endoral exposures in dental radiography.

The anode is made of tungsten only in the region of the focal spot and elsewhere of copper which dissipates the heat generated in the focal spot very effectively toward the outside to the vacuum housing surrounded by oil.

Rotating anode tubes

Depending on their technological construction, rotating anodes can be loaded up to 100 kW due to their rotation. Due to the very high thermal loading, rotating anodes are designed as compound anodes, i.e., they are made of different materials or compounds of elements that have proven to be particularly effective in generating radiation and at the same time meet the high physical requirements.

The rotation of the anode significantly increases its loadability in comparison with a stationary anode tube. The surface available for heat absorption is increased by the factor of the focal spot width times the focal spot loop as a result of the rotation. This geometrically enlarged surface consequently multiplies heat absorption and heat emission.

Fig. 7.13 illustrates the basic structure of a rotating anode. The anode disk is made up of one or two support disks (1 or 1 and 2) which may be different thicknesses (masses) according to the required heat dissipation (see pictures

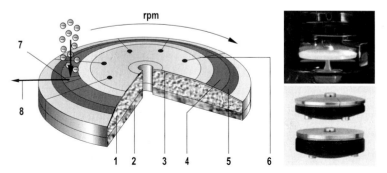

Figure 7.13 Basic diagram of a rotating anode

7 X-ray System Components

to the right in the figure). The anode disk diameter is 100 mm, 120 mm or 125 mm and also depends on the required type of performance.

Disks with only one support have a forged molybdenum disk (1) of 4 mm to 19 mm depending on the design. This disk contributes to increase heat capacity (heat storage capacity) and decreases crack formation of the anode.

The tungsten-rhenium alloy used for the anode surface first achieves good elasticity in a temperature range of 100 to 200°C. Prior to that, it is brittle. When releasing the first exposure (when the anode is still "cold") with a high tube current I_R (mA) and high exposure voltage U_R (kV), extremely high temperature increases occur (temperature curve) within the anode disk. This generates high mechanical stresses which can result in the formation of cracks. To prevent such heat cracks, thin radial slots (Fig. 7.13-6) are produced in different anode disk designs (comparable to the expansion joints of railroad rails, bridges, or buildings). The X-ray tube is then referred to as an X-ray tube with a "relaxed" rotating anode.

High-performance tubes have a second graphite support layer of approximately 9 mm (Fig. 7.13-2). The top part of the disk is the radiation-generating layer of approximately 1 to 1.3 mm (3) of tungsten or an alloy of tungsten (90%) and rhenium (10%).

Figure 7.13 shows an example with two offset focal spot paths at different inclination angles (4 and 5) and also shows the electronic focal spot (focus, 7). Offset focal spot paths generate a so-called focus flaw (see chapter 9).

The central ray, i.e., the reference axis of the useful beam (8), is only geometrically indicated. Physically speaking, the result is an indifferent X-ray spectrum, which is collimated to form a geometrically useful beam by the housing exit window and particularly by the primary or multileaf collimator.

7.3.4 Focal spot, focus

According to the number of spiral-wound filaments in the cathode, the same number of focuses with different geometrical dimensions can be generated on the anode. The focus size is selected according to the requirement for optimal radiographic visualization (see section 9.3 on focus sizes, focus unsharpness).

Stationary anodes only have one focal spot (focus) while rotating anodes can have up to three focal spots (focuses). These are referred to as 1, 2 or 3 focus tubes.

Figure 7.14 shows that the electrons (2) emitted by the spiral-wound filament impinge on the rotating anode (1) in a relatively focused manner within a defined region, and form the focal spot or focus for the radiation beam created there. The following differentiations are made:

7.3 X-ray Tube

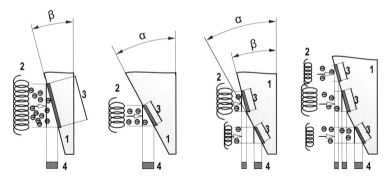

Figure 7.14 Schematic diagram of the line focus principle

- Electronic focal spot;
  the size of the electronic focal spot is determined by the width of the electron beam emitted by the spiral-wound filament (the cathode) and impinging on the anode in the direction of the longitudinal axis of the rotating anode (Fig. 7.14-2).
- Thermal focal spot;
  the angle of inclination of the anode (α, β, etc.) results in a greater value referred to as the thermal focal spot (Fig. 7.14-3).
- Optical focal spot or focus;
  the outgoing cone of radiation is theoretically dependent on the angle of inclination of the anode (α, β, etc.) and the size of the electronic focal spot referred to as the optical focal spot (Fig. 7.14-4).

2 and 3-focus tubes have different focal spot sizes. The following designs are available:

- Two adjacent spiral-wound filaments.
  Only one focal spot path is used for both focuses. As a result, only one anode angle of inclination is possible (no focus flaw).

Focal spot/focus designation	Max. permissible dimensions (mm)	Anode angle	Usable format for SID = 100 cm
0.2	0.30 x 0.30	6°	20 cm x 20 cm
0.4	0.60 x 0.85	10°	33 cm x 33 cm
0.6	0.90 x 1.30	12°	41 cm x 41 cm
1.0	1.40 x 2.00	17.5°	58 cm x 58 cm

Figure 7.15 Examples of the size designations of focal spots and the corresponding real dimensions (left) and the usable radiographic surfaces as a function of the anode angle (right) (see section 9.3.3)

7 X-ray System Components

- Two consecutive spiral-wound filaments. One focal spot path is available for each focus. Two different anode angles of inclination allow different geometric exposure relationships.
- Three-focus tubes use a combination of two adjacent spiral-wound filaments and a single spiral-wound filament for electron focusing.

The size of the focal spot affects the achievable image sharpness. The larger the focal spot, the greater the loadability of the anode but also the greater the geometric unsharpness (see section 9.3).

Anode angles affect the size of the usable surface of the image to be recorded as a function of the distance from the focus (focal spot) to the image receptor plane (source-image distance) (see Fig. 7.15, section 9.4). Focal spots have a maximum allowed size. Designation of the focal spot values and the assigned dimensions are standardized according to DIN and IEC (see DIN and IEC in the appendix).

7.3.5 Rotor and stator

The rotor, also referred to as the "runner" in electrical engineering, is the rotating part of an electrical machine (an electric motor). The anode disk is connected to an axis, this rotor. To bring this rotor including the anode disk to the necessary rotation speed, a magnetic field is rotated in an annular manner around the rotor so that its poles form a circular path. This generates a magnetic rotating field. This magnetic rotating field is generated by a stationary, annular stator around the rotor, also referred to as a "stand" in electrical engineering, and three coils which are offset by 120° and through which a three-phase alternating current flows.

The number of revolutions for the rotor depends on the alternating current frequency or the alternating current frequency of the three phases supplied by the rotating-anode starter in the stator. This is normally 150 Hz or 180 Hz.

In the case of most tube types, the rotor rotates on ball bearings connected to the tube housing. In the case of tubes under significant load, particularly in cardiac angiography and interventional angiography workstations, the rotor pivots in "liquid metal" (see section 7.3.1, Fig. 7.8).

7.3.6 The rotating-anode starter

To extend the service life of an X-ray tube (without liquid-bearing technology), the rotating anode is accelerated to the necessary number of revolutions prior to every exposure in order to protect the ball bearing and to reduce the total mechanical load.

Regardless of the ball bearing or liquid-bearing technology, the number of revolutions n of the anode can be 1,200 to 9,000 rpm and, for special tubes, up to 17,000 rpm, depending on the tube design and the anode drive frequency. The anode speeds result from the speeds of the stator magnetic field rotating around the rotor as a function of the alternating current frequency (see the rotor section).

During fluoroscopy, the anode rotates at a continuous drive frequency of 20 Hz with 1,200 rpm. For startup times, a rotational speed of 2,800 rpm results in approx. 0.8 seconds for X-ray images with a rotation driven by 50 Hz, 8,500 rpm in approx. 1-2 seconds for 150 Hz, and an anode speed of 17,000 rpm in approx. 2.5 seconds for 300 Hz.

In contrast to rotors with ball bearings, the anode disk rotates continuously even during exposures (CAT = continuous accelerated tube) when using liquid-bearing technology with liquid metal lubrication. The anode disk no longer needs to be accelerated prior to every radiation release. It is started or stopped when the X-ray system is switched on and off and maintains a continuous number of revolutions n of 9,000 rpm at a 150-Hz anode drive frequency during the entire period of activity.

7.3.7 Other processes in the X-ray tube

In addition to the desired generation of X-rays, other physical processes that are not desired but are not preventable also result in the X-ray tube.

Extrafocal radiation

Extrafocal radiation is scatter radiation resulting when part of the electrons emitted by the cathode are scattered back by the anode (without energy loss). These electrons reach the potential gradient between the cathode and the anode and impinge again on the anode outside of the original focal spot (focus). This generates low-energy X-ray quanta. The percentage of extrafocal radiation can be up to 20%. Appropriate constructive tube design measures, e.g., influencing of the electrical field distribution in the critical region of the anode or using collimators near the focal spot, can reduce the extrafocal radiation to 3%. Extrafocal radiation affects the image quality with a loss of contrast.

Stem radiation

Stem radiation is extrafocal radiation not originating from the focal spot (focus). It occurs when reflected electrons impinge on the anode stem (outside of the focal spot) and generate quanta there. It forms the majority of the extrafocal radiation.

7 X-ray System Components

Leakage radiation

The primary radiation generated in the focus, including the extrafocal radiation, travels in all directions within the tube housing. X-ray quanta penetrate the vacuum housing of the tube, albeit in a weakened state. Therefore, the tube is installed in a protective lead tube housing. Despite this lead shielding, a residual remainder penetrates the protective housing. This remainder is referred to as leakage radiation. For data sheet information, the leakage radiation is provided per time unit (hour): 0.7 mGy/h. The allowable limit value is 1 mGy/h, and the measured values generally yield smaller values unless the X-ray tube is defective.

Heel effect

The heel effect refers to the single-sided drop in dose intensity of x-ray quanta, in the useful beam at the anode (Fig. 7.16-3). The intensity of the radiation (X-ray quanta) decreases as its exit angle (α) to the anode surface decreases.

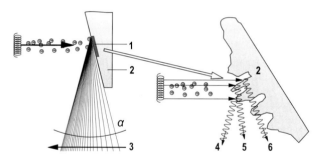

Figure 7.16 Schematic diagram of the heel effect. The roughened focal spot shows how the X-ray quanta are partially still weakened within the anode material.

Since the quanta were partially generated under the anode surface (penetration depth), they travel a greater path within the anode when the exit angle (α) to the anode is small and are absorbed more strongly there. Figure 7.16 on the right shows the generation of X-ray quanta in a roughened anode material:

- Quanta (6) occur in a crack and are repeatedly weakened within the anode since their radiation angle α to the anode surface is very small.
- Quanta (5) occur in a crack. However, they are weakened within the anode less than the quanta (6) as a result of the greater radiation angle α.
- Quanta (4) occur on the anode surface and have a greater radiation angle α. They are not weakened within the anode.

Quanta 5 and 6 illustrate the heel effect. This otherwise undesired effect is useful in mammography as a result of the anatomical shape of the female breast.

Inherent filtration

Inherent filtration refers to the absorption (filtration) of low-energy radiation by the X-ray tube, i.e., by the wall of the tube receptacle, the coolant, oil, the X-ray window, etc. It is provided as Al equivalent values and is required by the authorities to have at least 2.5 mm Al equivalent total filtration (inherent plus additional filtration) (see section 4.1).

7.4 Primary Collimator

Fig. 7.17 illustrates the basic structure of a multileaf collimator. The radiation issuing from the X-ray window (1) of the X-ray tube is referred to as the primary radiation. A primary collimator or a cone encased in lead, e.g., for dental X-ray tubes, must be attached to every X-ray tube to laterally limit the primary beam.

The portion of the primary radiation actually used for exposures and fluoroscopy, i.e., the radiation collimated by the cone or the primary collimator, is referred to as the useful beam.

Collimation of the useful beam field to the smallest possible area is a very important measure for keeping patient exposure as low as possible. Moreover the smaller or more exact the collimation of the useful beam field, the

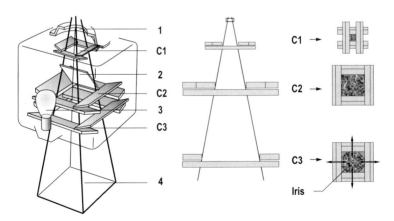

Figure 7.17
Schematic diagram of X-ray collimation by the collimator or primary diaphragm

less scatter radiation (image quality). Collimation is performed via manual, motor-controlled, or automatic movement of the collimator leaves or plates. They are attached near the focus and near the object (Fig. 7.17-C1 through C3) and synchronously move the sides of the exposure field with respect to height and width.

Primary collimators (Fig. 7.17) are differentiated into:

- Multileaf collimator: A primary beam collimator with several consecutive collimator systems positioned in the "depth" field (C1 through C3).
- Double-slot diaphragm: A primary beam collimator with two lead plate pairs on one plane (e.g., only C1) for the lateral length of the exposure field for each pair. Double-slot diaphragms are primarily used for mobile generators.
- Iris collimator: In image intensifier radiography, collimation is performed via quasi circular collimation of the image intensifier (I.I.) entry diameter using an iris collimator (C3 and iris).

For radiation-free pre-collimation or to make the expansion/restriction of the radiation field visible prior to exposure, i.e., without radiation, a halogen light (Fig. 7.17-3) is installed with a timer in the collimator. This is referred to as a primary or multileaf collimator with a light localizer, light-beam diaphragm, or light localizer collimator. A mirror (2) directs the light field collimated accordingly by the plates to the object to be recorded. Different systems have a laser light localizer which displays the center of the useful beam cone with two crossed thin laser light lines. Laser light localizers are primarily used in surgical C-arm systems or when precise positioning of the object of interest is important.

When the collimator is completely covered, e.g., in the case of angiography systems (Fig. 7.18-3 and 4) or in case of undertable X-ray systems, radiation-free field collimation cannot be performed using a light localizer. The size of the collimated field is typically displayed in centimeters or inches. The size and position of collimation can also be set or displayed on the last fluoroscopic image on the monitor (last image hold, LIH). This requires a digital imaging system (see chapter 9).

Examples in Figure 7.18-1 through 4 show how the collimator can be generally mounted upstream from the X-ray window of the X-ray tube. Based on

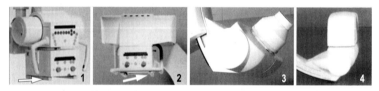

Figure 7.18 Examples of primary collimators with tube

the example of an angiography system (3 and 4), the X-ray tube is connected to a rotatable collimator. Both units are enclosed by a customized hygienic covering to protect against damage. Figure 7.18-3 shows a typical covering for a multileaf collimator for cardiological examinations. It includes plate or leaf forms especially designed for collimating the heart. Figure 7.18-4 shows a rotatable multileaf collimator in an angiographic radiographic system. As a result of the rotation of the multileaf collimator, the object is always displayed on the monitor in the correct position.

Special filters, a measurement chamber for measuring the dose area product, or other collimator accessories can be inserted in the accessory rails of the collimator (Fig. 7.18-1 and 2) for the purpose of radiation hygiene.

7.5 Spotfilm Device and Bucky Diaphragm

Spotfilm devices, cassette trays, bucky trays, catapult bucky trays or bucky diaphragms (see appendix for bucky) are system components for image receptor system exposures (see chapter 8).

The term "spotfilm" comes from "spotfilm radiography." This refers to an imaging system capable of triggering an X-ray in a "targeted" manner by presetting the region of interest under fluoroscopy.

A spotfilm device or cassette tray is necessary for direct radiography in which exposures are created on cassettes using X-ray film or imaging plates. To trigger an exposure, the cassette is inserted from a park position (from outside the useful beam field) or via manual insertion in fractions of seconds into the exposure position and is then returned to the park position following exposure or is ejected again and manually removed.

Different size cassettes can be inserted in these spotfilm devices (Fig. 7.19). In the case of all spotfilm devices, the cassette is positioned symmetrically

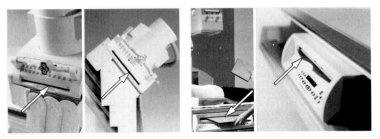

Figure 7.19 Spotfilm device: on an undertable (left) and on an overtable (right) radiography system. The cassette is inserted in the cassette tray of the spotfilm device (arrow) and positioned accordingly.

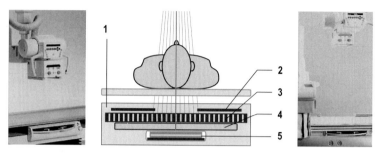

Figure 7.20 Basic structure of a spotfilm device or a catapult bucky tray (center) and catapult bucky tray of an X-ray patient table (left) and a UT fluoroscopy system, e.g., for a second exposure plane (right)

with respect to the central ray and collimated near the film, similar to the primary collimator, to the desired image size using lead plates. Divisions allow a number of exposures adapted to the particular object to be made on the same film (see chapter 8 on cassettes).

C-arm fluoroscopy systems with image intensifiers, as used in angiography or surgery, do not have a spotfilm device. However, X-ray film/imaging plate exposures can be made using cassette holders for cassettes and stationary grids positioned upstream from the I.I. input screen.

As a result of advancing digitization in imaging systems using image intensifiers and CCD technology, or in imaging systems with flat detectors, the term "spotfilm device" is no longer apt but the century-long designation of this system component is still too engrained in radiography. Every desired "spot film radiography" can also be digitally stored during fluoroscopy.

Catapult bucky trays, cassette trays, bucky trays or cassette bucky trays, or those generally referred to as bucky diaphragms, are system components similar to the spotfilm device. These are primarily used in X-ray systems that do not require any fluoroscopic examinations. The different designations have resulted from different "contents" or functions.

A bucky or catapult bucky tray (Fig. 7.20-1) includes:

- A receiving and clamping device for receiving cassettes with a film-screen combination, an imaging plate or a flat detector (5). When using a flat detector, this is also referred to as a detector tray.
- A stationary or movable grid for reducing the scatter radiation on the image receptor system (3).

The term catapult bucky tray is correct when the grid for reducing the scatter radiation is moved during the exposure (in a "catapult-like" manner) (see chapter 10 on grids).

When using multi-line grids, it is not absolutely necessary to move the grid. The position of the grid remains constant during the exposure. In this case, bucky tray or cassette bucky tray is appropriate. In the case of image acquisition systems without a grid, the designation cassette tray is correct. Cassettes with an integrated stationary grid known as grid cassettes are typically used.

Since the introduction of the honeycomb grid by Gustav Bucky, the term bucky diaphragm has become established for imaging devices with a grid. Radiographic workstations with a bucky diaphragm, i.e., with a grid device for reducing scatter radiation to the image receptor medium, are also referred to as a bucky table, bucky wall stand or bucky workstation.

For a high-quality X-ray image with radiation and scatter radiation protection, the following components are included in the spotfilm device or catapult bucky tray:

- A dose measuring chamber (Fig. 7.20-4) for automatically switching off the radiation when the dose required for the correct exposure is reached (see section 4.3 and chapter 9).
- Lead plates for additional collimation of the useful beam field near the film onto the image receptor medium (Fig. 7.20-2). This collimation reduces the scatter radiation behind or after the object to the peripheral zones of the imaging field.

7.6 Stands

Practically no radiography system is designed without some form of stand. Stands are system components for supporting imaging systems, X-ray tubes and collimators or monitors, etc. Therefore, it is always important to take the function of the stand into consideration in the designation. They should always be described fully according to their purpose whenever misunderstandings may occur.

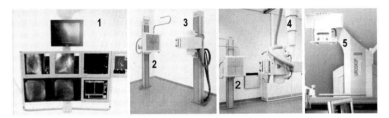

Figure 7.21 Ceiling-mounted monitor or display support stand (1); floor-mounted bucky wall stand (2), floor-mounted (3) and ceiling-mounted (4) tube stand. Integrated tube system stand (5); see the illustrations in section 6.3.

7 X-ray System Components

The term "stand" does not specify the use or application area. Stands may be described according to their location, such as:

- Ceiling-mounted stand
- Floor-mounted stand
- Wall stand
- System stand (integrated, stationary or mobile)

However, it is still not clear what task is assigned to the particular stand. Therefore, they must also be described according to their radiological use or radiological "task", such as:

- Tube stand
- Vertical bucky stand
- Monitor or display support stand
- C-arm stand, etc.

Overtable fluoroscopy and urology systems have a tube stand connected to the basic system, and the X-ray tube coupled to the collimator is rotatable for oblique and horizontal projections.

Undertable fluoroscopy systems can be equipped with a second exposure plane in addition to the spotfilm device. The installation of a catapult bucky tray and a second exposure plane, i.e., a second X-ray tube with a collimator, is necessary for this purpose (see Fig. 7.20, right). This second X-ray tube coupled with a collimator is attached to a ceiling-mounted tube stand. This makes possible vertical (a.p./p.a.), oblique, lateral and conventional tomography on the basic system as well as oblique and horizontal exposures on a floor-mounted and/or wall-mounted vertical bucky stand.

Of course, this second exposure plane can be set up in connection with urological or universal overtable fluoroscopy systems as well as bucky/grid patient tables.

Self-supporting X-ray systems consisting only of a vertical bucky stand, for example, must be equipped with a radiation-generating system. X-ray tubes and collimators can be attached to a ceiling-mounted tube stand or a floor-mounted tube stand in this context.

Monitors or displays, typically with fluoroscopy and angiography systems, are installed in examination rooms, preferably in ceiling-mounted support systems designed for this purpose. The terms monitor or display support system (MSS or DSS) have now mostly replaced the term "monitor support stand."

Designations, such as ceiling-mounted or floor-mounted C-arm stand, were also used for C-arm systems in angiography or surgery. This refers to the stand as a support for the C-arm.

7.7 Accessories

Proper accessories and the correct use of these aids support physicians and their teams as well as the patient during radiological examinations or interventional procedures. Accessories help to ensure safety during patient positioning, and improve radiation hygiene and image quality.

Accessories are all devices that are not permanently attached to the X-ray system but are individually attached as needed to the appropriate location. They can be positioned on the patient table, stand on the floor, or hang from the ceiling; they can also be inserted in the accessory rails of the collimator or patient table.

Accessories are an important part of all radiography systems. They are indispensable for careful, safe radiography, as well as maintaining radiation hygiene and improving image quality. Therefore, all manufacturers provide comprehensive accessory catalogs and data sheets.

The accessories shown in Figure 7.22 can be divided into different areas of application.

7.7.1 Patient safety

Accessories like protective head-end rails (Fig. 7.22-6) and side rails (Fig. 7.22-1), as well as hand grips (Fig. 7.22-3), give patients a sense of security, and support comfortable positioning. They are used for head-down positions up to approximately 10° to 15°. For a 15° or more head-down position, shoulder supports must be used to keep the patient from slipping (Fig. 7.22-5); for

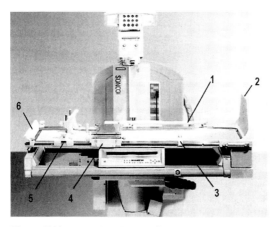

Figure 7.22 Examples of an OT fluoroscopy system showing some of the many uses of X-ray systems

7 X-ray System Components

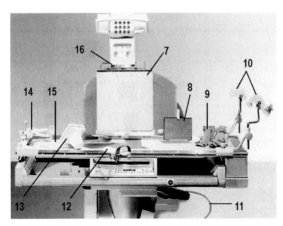

Figure 7.23 Examples of an OT fluoroscopy system showing some of the many uses of X-ray system

a head-down position of more than 40°, ankle braces must be used (Fig. 7.23-9). Examinations from the horizontal to vertical position always require a footboard (Fig. 7.22-2).

7.7.2 Patient comfort

To make examinations and interventional procedures more pleasant for patients as well as for physicians and their teams, a number of positioning aids and mattresses should be available (Fig. 7.23-15). Secure and comfortable positioning, which also takes anatomical requirements into consideration, supports comfortable work and accelerates the workflow. In the case of angiographic examinations, the arm positioning (e.g., for IV, injection, etc.) is supported by special devices (Fig. 7.23-12).

Skull and cervical spine examinations are supported by special, three-dimensionally adjustable head supports (Fig. 7.23-14). Leg supports are used for gynecological or urological examinations (Fig. 7.23-10). As a result of cassette holders (Fig. 7.23-8), the patient does not need to be repositioned for lateral exposures, and a foot switch for fluoroscopy and exposure (Fig. 7.23-11) supports the physician during examinations close to the patient.

7.7.3 Image quality and radiation hygiene

To improve image quality, density differences are compensated for by using transparent compensators, special preliminary filters or compression devices such as a compression cone (Fig. 7.23-13) or a compression band (Fig. 7.22-

130

4), overexposure at object transition regions are compensated for, motion artifacts are minimized or eliminated, and scatter radiation is reduced.

Special devices directly at the X-ray system protect against unnecessary exposure to radiation and also facilitate an ergonomic examination workflow. Dose measuring devices can be inserted in the accessory rails of the multileaf collimator (Fig. 7.23-16).

7.7.4 Synchronous technique

IV bottle holders, EKG cables, etc., can be attached directly to the patient table so that the corresponding accessories move with patient movement.

7.7.5 Hygiene and sterile work area

Cleanliness and hygiene are important aspects for the patient, the examiner, and the electronic or mechanical components. Therefore, sterilizable fabric covers or transparent, sterile disposable covers in different sizes and shapes are available for the different system components.

7.7.6 Pediatrics

Radiography systems for special pediatric examinations are no longer produced. Therefore, special accessories are very important for children and infants. Particularly the positioning of smaller patients on the "large" radiographic systems requires sensitive equipment.

Figure 7.24 shows a positioning shell (BABIX shell) with a support for vertical positioning of small patients for exposures on a vertical bucky stand (1) and for horizontal positioning on an X-ray patient table (2).

Using a positioning cradle that is rotatable about the longitudinal axis and is attached to the accessory rails of the patient table (3), exposures can be completed with virtually oblique projections (see section 9.4.1).

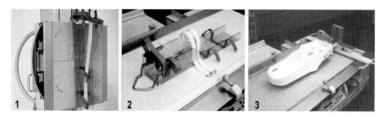

Figure 7.24 Pediatric positioning aids

8 Image Receptor Systems

Breakthroughs in technology have brought vital changes to radiological imaging, especially in the field of digital imaging. However, the emergence of digital imaging systems does not signal the demise of conventional methods, such as X-ray film and the image intensifier. Instead, digital technology is adding to and extending the diagnostic capabilities of these systems. Nonetheless, high-tech standards and ongoing development of digital systems based on flat detector technology point to a decline in the near future in the percentage of radiological exams that use X-ray film, image plates and image intensifiers.

Image receptor systems are differentiated according to their imaging technology:

- Film-screen systems (analog radiography)
- Image plates (DLR or digital luminescence radiography)
- IITV (image intensifier television) systems (analog and digital fluororadiography)
- IITV systems with CCD sensors (digital fluororadiography)
- Flat detectors (digital radiography and fluororadiography)
- Detector systems for computed tomography (digital radiography)

8.1 Image Modes

Certain disciplines require single static images only, e.g., orthopedics, internal medicine to some extent, and chest and gastrointestinal studies in general. Other studies require series images, or the cine method, of fast dynamic processes like swallow studies, cardiac function, vessel functioning, etc.

8.1.1 Analog and digital

The first digitally generated images were created with the aid of computers in the early half of the 1970s. By converting X-rays into analog voltages, amplifying and digitizing them, and then using the computer to reconstruct the data, the first computed tomography (CT) images that could be used for diagnosis emerged.

The introduction of the image plate as a digital imaging medium in the first half of the 1980s launched the first serious foray against film-screen systems. Image plates lost some of their edge, however, when flat detectors came on the scene in early 2000.

Digitizing analog image signals in the image intensifier chain began around 1988 with fluoroscopy systems.

The first image intensifier television systems came into use around 1989, applying the CCD sensor technology already established in entertainment electronics (TV/video cameras). This digital image acquisition medium increasingly supplanted the conventional TV image tube.

Flat detectors (also called solid-state detectors or flat panel detectors (FPD)) as image acquisition systems made the definitive breakthrough in digital imaging in 1999. Beginning with single-shot X-ray workstations, this technology gradually overtook the entire spectrum of radiography systems in use today, as described in chapter 6.

8.1.2 Direct, indirect imaging

The differences between imaging systems lie in their technical or technological solution. The terms "direct" and "indirect" imaging have a historical basis. At first, X-ray film was considered the medium of direct radiography, even though light was and is chiefly used to expose it.

When image intensifier television technology was introduced, the radiography was viewed indirectly. In the IITV chain, X-rays were converted into light and electrical signals and the image then displayed on a monitor. Since digital images arose by converting X-rays into light and electrical signals, which were then digitized, these images were also categorized as indirect.

In the field of flat detector technology, however, the terms "direct digital" and "indirect digital" have come into use. Direct flat detector systems convert the X-rays directly into electrical charges via a layer of material sensitive to radiation (e.g., amorphous selenium, a-Se). In indirect flat detector systems, X-rays first generate visible light in a scintillator (e.g., cesium iodide, CsI, like the image intensifier). Light sensitive photodiodes then convert the light into electrical charges.

8.2 Film-screen Systems

Except for the early fluorescent screens, film and film-screen images were the original method for generating a radiograph for diagnosis. X-ray film and film-screen images continue to play a distinct role, even in today's digital world.

8 Image Receptor Systems

Figure 8.1 X-ray image from the pioneer days (left): the patient had to hold the film cassette himself. X-ray image on a modern floor-mounted X-ray image stand with catapult bucky tray (right).

For this reason, film-screen systems will be thoroughly discussed here.

Film-screen systems are viewed as the combination of conventional X-ray film and intensifying screens. X-ray film is exposed almost exclusively to light and intensifying screens are used to enhance film exposure. Intensifying screens have a light-sensitive layer made of highly fluorescent phosphors. X-rays are converted into light according to their intensity, and the light in turn exposes the film.

Using intensifying screens considerably reduces the patient's exposure to radiation and markedly improves image quality as required by radiological specifications and radiation protection laws.

Today, X-ray film and film-screen systems (and imaging plates) are housed in tightly sealed cassettes of aluminum or plastic that are highly transparent to X-rays but keep out ambient light.

8.2.1 The X-ray film cassette

From the outset, X-ray film was placed in and removed from the cassette in the darkroom and inserted into the film developing system. Daylight systems that automatically control workflow are taking over darkroom work. Daylight systems, however, are not able to support all the cassette sizes in use today, so darkrooms are still in use.

There are different cassette sizes for X-rays of all body parts and organs, from head to toe: from 13 cm × 18 cm (5" × 7") to 35 cm × 43 cm (14" × 17"), aligned with or transverse to the longitudinal axis of the patient. Depending on the X-ray system's image program, several sub-sections of a film can be exposed individually, e.g., as a series, in a process known as film sectioning. Manufacturer data sheets and user manuals usually contain graphical representations of the film sectioning options of their cassette programs. X-ray film cassettes are standardized by the IEC, DIN and ANSI (see appendix).

8.2 Film-screen Systems

 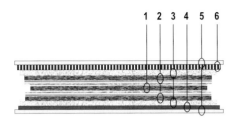

Figure 8.2 Inserting a cassette in a cassette tray, and schematic diagram of cassette with a film-screen system

Depending on the application and type of X-ray system, film cassettes are positioned by hand or automatically (in an image receptor or cassette tray) at the site of the image. Strict radiation protection procedures have to be followed.

Fig. 8.2 shows: 1) X-ray film; 2) intensifying screens, with front and back screen; 3) pressure pad, e.g., plastic foam that is glued to the cassette plate; 4) lead foil insert; 5) top and bottom of the cassette, made of aluminum or plastic; some cassettes have a stationary grid (6) and are called grid cassettes.

Handling and quality

Although the cassette's "only" job is to keep out light and maintain good contact with the intensifying screen(s), the importance of the cassette with regard to image quality should not be underestimated.

One potential weak point in the cassette is slackening pressure of the pad against the film. At points where the pressure is weaker, the intensifying screen(s) and X-ray film are not quite flat and tight against one another. This causes additional scatter radiation and therefore blurring on the X-ray, also known as *cassette unsharpness*.

There are two ways to test whether the contact over the entire film/screen surface is even; take an x-ray image using a test grid of fine wire or plastic mesh or a testing plate with specially aligned bore holes. The test grid or plate is placed directly on the cassette and a test image taken. Out-of-focus areas in the mesh structures or bore holes can indicate points of poor contact.

When film/screen contact has to be especially good, vacuum cassettes are used. The vacuum cassette contains a wall to apply pressure, creating a vacuum between the X-ray film and the intensifying screen.

Cassettes cannot be manufactured to a specified robustness. The front wall needs to absorb very few (none, if possible) X-ray quanta and therefore must be thin, typically 1 mm of aluminum.

Cassettes also need to be lightweight. Yet the back wall has a thin lead insert, usually 0.1 mm thick. The lead insert prevents any additional exposure of the X-ray film by scatter radiation from the back of the cassette frame.

In spite of their top-grade materials, cassettes are delicate and have to be handled appropriately. They should not be forced open or closed; the film must be carefully loaded and removed; and cassettes should not be dropped on the floor or stacked on top of one another, but stored in specially designed shelves.

8.2.2 Direct images with film changer and magazine technology

Continuing digitization of X-ray technology superseded a number of imaging techniques. Such systems are no longer manufactured or sold. As they are still found in older X-ray systems, however, a brief description follows.

Film changer for angiography

In the field of general and special angiography, for cerebral and peripheral vessel images, etc., the PUCK sheet film changer with a frame rate of 4 frames per second was used. There were up to 20 different programs for image series that could be selected, such as stepping, lowering the kV, etc.

The "see-through technique" was the name given to the ability to track the flow of contrast agent via the IITV chain behind the PUCK.

This technique required buffering, i.e., every image was temporarily saved in digital form so that it could be continuously displayed on the monitor during pauses between images (see chapter 9, gap filling).

Figure 8.3 Left: PUCK film changer for direct image technology with storage magazine for up to 30 film sheets, 35 cm × 35 cm in size. Center: PUCK program selector. Right: PUCK on an angiography system with schematic diagram of the beam path using the "see-through technique."

8.2 Film-screen Systems

Magazine devices for thoracic images

To accommodate very high numbers of chest X-rays taken daily, it became necessary to use systems that were equipped with a film magazine with approximately 50 large-format film sheets (Fig. 8.4). A large number of exposed films were assembled in a film magazine.

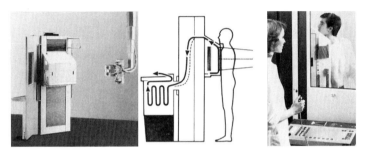

Figure 8.4 Chest imaging system with magazine technology and developer attached

There were also film magazines for orthopedic imaging tables and universal fluoroscopy systems. As system technology progressed, magazines fell out of use.

8.2.3 X-ray film

The quality of the X-ray image depends on the patient's condition, the parameters of the radiation-producing systems, the geometric factors and the practical application of the imaging technology.

The recording medium itself, the film, has considerable influence on the image quality, along with its configuration and appropriate care and use. This includes

- Film type
- Screen type
- Film or film/screen contact (flat and level in the cassette)
- Film development
- Film storage (safekeeping)

Figure 8.5 shows the structure of an X-ray film with two emulsion layers: the base (1), which supports the emulsion, is usually made of polyester. The adhesive layer (2) is used to attach the emulsion layer (3) to the base. The emulsion is a light-sensitive, photographic layer with light-sensitive silver halide suspended in gelatin. The layer is approx. 0.03 mm to 0.04 mm thick and the silver content is approx. 5 grams/m^2. The protective layer (4) is

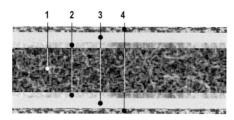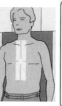

Figure 8.5 Schematic diagram of X-ray film with two emulsions and two intensifying screens. X-ray image thoracic spine, a.p. standing.

applied to guard against damage, prevent electrostatic charge and improve sliding.

Emulsion

Since X-ray film is basically photographic film, its sensitivity to light is considerably higher than to X-rays, approx. 95% to 97%. Only about 3% to 5% of the X-ray quanta is absorbed by the emulsion layer of the film. Therefore, the emulsion or emulsion layer is sensitized to the light emitted by the intensifying screens. The layer carries the data for the latent X-ray image. Emulsion layers are made of silver compounds or silver halide suspended in a gelatin medium to keep them evenly dispersed. See the appendix for information on halides.

The silver halides evenly suspended in the gelatin matrix belong to a group of chemical compounds of silver and chlorine, bromide or iodine. When silver reacts chemically with chlorine, the result is silver chloride (silver chloride crystals); when silver reacts with bromide, silver bromide (silver bromide crystals) is produced; and in reaction with iodine, silver iodide (silver iodide crystals) results. Silver halides are extremely light-sensitive crystals, which is why they are commonly used in photography.

Since the photographic emulsions absorb only very few X-ray quanta, the layer is relatively thick for screenless film (approx. 0.08 mm and 20 grams/m^2 silver content) and the required dose (quanta, intensity) is several times higher than for film-screen combinations.

X-ray film coated on one side only (single-layer film) is used with a single intensifying screen. The layer of emulsion is variably thicker than for double-layer film. Film-screen combinations with only one intensifying screen, such as used in mammography or for multiformat cameras, are being increasingly replaced with digital imaging technology.

When two intensifying screens are used, the film must be coated on both sides; the emulsion layer is typically 0.03 to 0.04 mm thick and contains 5 grams/m^2 of silver.

For the most part, X-ray film with two layers of emulsion are used. The X-ray film lies in the cassette between the two intensifying screens, a front and back screen (a screen pair); see Fig. 8.5.

Screenless film, which is used without intensifying screens, does not exhibit any screen unsharpness (see section 9.3) and therefore has excellent resolution. This type of film is used primarily in dentistry for endoral images.

Optical density (OD)

The characteristic curve, blackening curve, blackening, film curve, gradation curve or density curve all refer to the same thing: they describe the relationship between film exposure during X-ray (impinging intensity, dose) including handling of the film, such as development, storage, etc., and the resulting visible optical density (density range).

The optical density interprets the reproduction characteristics (quality) of the X-ray film based on the characteristic curve (Fig. 8.6). It is defined as a base-10 logarithm, since the human eye also sees logarithmically. See the appendix for information on logarithms.

The optical density D or OD is also defined as the logarithmic ratio of the incident intensity I_0 (95-97% light intensity, 3-5% X-ray quanta = 100%) to the light intensity transmitted by the film I_1 (%):

- D (or OD) = $\log I_0 / I_1$ = $\log 100\% / 10\%$ = $\log 10$ = 1
- D (or OD) = $\log I_0 / I_1$ = $\log 100\% / 1\%$ = $\log 100$ = 2

Example:

If 1/10 of the impinging light is transmitted, then the OD = 1; if 1/100 of the light is transmitted, then the OD = 2.

- Therefore, as optical density increases, light transmissibility decreases.

Fig. 8.6 is a schematic diagram of a typical characteristic curve. The gradation (γ, G) is the straight-line portion in the mid-section of the curve where the slope does not change. At an exposure log B = 0, the inherent veil (A portion) of the film indicates the film's optical density.

The toe (B portion) of the characteristic curve is the area of underexposure. The shoulder (D portion) of the curve is the area of overexposure.

Both areas B (image is too light) and D (image is too dark) are unsuitable for radiographic purposes because of loss of contrast.

γ is the gamma value (G) and describes the steepest slope of the characteristic curve. The straight-line portion of the curve (C portion) describes the average gradient. That is the mid-section of the usable density curve, i.e., the optimum region of the film to expose depending on the contrast properties desired.

8 Image Receptor Systems

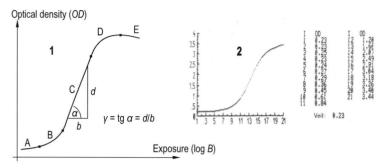

Figure 8.6 A typical density curve (1) and display on the PC monitor (2)

The E portion is the solarization portion (the reversal of the effect of light when film is highly overexposed); different X-ray positive films work in the solarization portion.

The shape of the optical density curve is determined by test-exposing an X-ray film to various intensities (step wedge). The optical density curve is obtained based on the visible and measurable blackening steps.

Sensitometers are also used to expose the film. They expose a test strip, e.g., to 21 different intensity steps at the same time. A densitometer is used to measure the resulting optical densities. Connected to a PC, the automatic densitometers measure the 21 densities in just a few seconds. The result is displayed on the PC monitor (Fig. 8.6, right).

The gradations of exposure (of the individual blackening degrees) in this example show the exposure doubling every other step. This means that at exposure step 21, the exposure is approx. 1,000 greater than at step 1. For this reason, the optical density is depicted logarithmically in the abscissa of the coordinate system. See the appendix for information on logarithms.

Quality control for film processing

Quality assurance steps in X-ray diagnostics are regulated by law and include acceptance testing, regular constancy check of the X-ray system, as well as regular review of the X-ray film processing.

The density curve is not simply a property of the film; film processing also plays role. Only if the development process is optimal, e.g., the chemicals are in good condition, the temperatures and processing times are correct, will the optical density curve deliver the optimal contrast.

The quality of the X-ray image is also influenced to a large degree by such factors as developer temperature, development time and composition of the chemicals. There is a procedure for testing film processing, whereby after the test film is developed (sensitometer strips), the following values are obtained through densitometer measurements of the optical densities.

- Base plus veil (A):
  the lowest optical density D_{min} value where the density curve (characteristic curve, blackening curve) begins; see veil.
- Sensitivity index (B):
  the optical density value of the step just above the optical density of the veil, or: $D_{min} + 1.0$.
- Contrast index (C):
  the density 4 steps above the sensitivity index, or D_{min} + approx. 2.4.

The procedure is known as the three-point method. The development temperature of the film can be shown as well. See Fig. 8.7, right.

The basic idea underlying constancy testing in film processing is that it must be completely independent of the X-ray system. The test films are therefore exposed using a sensitometer. These sensitometers are used daily for only brief periods of time, which means demands on them are low and they usually perform consistently over long periods of time. The same sensitometer should always be used.

Sensitometers, densitometers

Sensitometers use a step wedge with 21 steps (21 blackening fields) to expose a test film to a standardized light source. The "wedge constant" is 0.15, the difference in density from one step to the next. This means that the difference of the logarithmic exposure between one step and the next is 0.15, and to the one after that, 2 times 0.15 = 0.3. Hence the exposure doubles every other step. See Fig. 8.7.

In order to be able to expose both blue and green-sensitive film, sensitometers have a "blue-green switch." See section 8.1.5 for information on blue/green sensitivity. The test X-ray film must be exposed in a darkroom or a dark space.

A densitometer is used to measure the optical density of each step of the exposed and developed film strip.

Fig. 8.7, bottom: The sensitometer strip was exposed using 21 different exposure steps. Step 3 received twice the exposure as step 1, step 5 two times more than step 3, or $2 \times 2 = 2^2 = 4$ times the exposure as step 1. At step 21, the exposure is $2^{10} = 1,024$ times greater than the exposure at step 1.

Fig. 8.7-1: Optical density of the sensitometer strip. The 21 points are the measurements of the 21 exposure steps.

A = D_{min} ; B = D_{min} + 1; C = D_{min} + approx. 2,4.

8 Image Receptor Systems

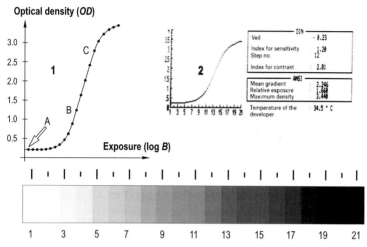

Figure 8.7
Automatic analyses of a sensitometer strip and display on a PC monitor (2)

Fig. 8.7-2: The automatic sensitometer measures the 21 density values and transmits them to a PC. Immediately the density curve and the DIN and ANSI analyses are displayed on the monitor.

Densitometers are instruments of simple construction. They basically consist of a constant light source, an opposite photocell to measure the impinging light, an amplifier to magnify the signal, and a digital display for the optical density. The light that impinges on the photocell and the resulting electronically amplified signal are used to measure the optical density, or blackening, of the X-ray film.

Veil

Every film has a minimum density or blackening value based on its composition. For this reason, the characteristic curve does not start with an optical density of 0, but somewhere between 0.16 and 0.24. This minimum value is known as the veil (basic veil) and should not exceed 0.25, because higher veil values reduce the contrast (Fig. 8.7-A).

Higher veil values can be caused by: film obsolescence (stored too long), exposure to dark room lighting or other light sources, defective packaging, or an error in film processing.

Gradation

The characteristic gradation, or blackening curve, indicates the range of optical density or density latitude of the X-ray film based on its optical density properties. Only the straight-line portion (Fig. 8.6-C) of the curve, the "average gradient", is usable for X-ray diagnosis.

The characteristic curve shows the ratio between the relative exposure log B (= brightness times exposure time, intensity measured as a base-10 logarithm) and the resultant optical density range (latitude).

The film with the steeper slope or rise in the straight-line portion of the curve (Fig. 8.8, left) creates greater optical density latitude, i.e., an image with considerably more contrast. The film with the flatter slope (Fig. 8.8, center) has less contrast, its density latitude is narrower, and the image has more gray areas. It is often referred to as "flat film."

The gradation of the film is not the only factor in determining the contrast of an X-ray image. Under the exact same conditions (object and exposure data), two sheets of film with different characteristic curves produce different images (image effects). In a similar fashion, look-up tables can produce different image effects when digital methods are used. See section 9.3 for information on look-up tables.

X-ray film sensitivity

The term sensitivity (light sensitivity) is borrowed from photography. In photography, sensitivity is defined internationally in ISO levels, e.g., 200/24° (see appendix). It expresses the amount of light needed over a defined period to reproduce all the shades of gray/color (after development) as perceived by the human eye. The sensitivity range of photographic layers (emulsions) extends from the γ rays into the infrared range (wavelengths from 0.0001 to 1,200 nm) (see Fig. 2.9).

The degree of sensitivity of X-ray film is also determined by the amount of light (intensity of the exposure) required to achieve an optical density that is

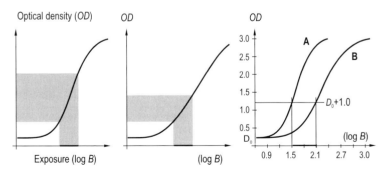

Figure 8.8 Schematic diagram of characteristic curves (left and center), see Fig. 8.6. and sensitivity comparison of two films. Film B requires approx. four times the dose of film A (right).

8 Image Receptor Systems

usable for diagnosis. It depends on the thickness and density of the object being imaged and therefore the intensity of the X-ray (required dose).

By exposing the film using the sensitometer and then measuring the blackening steps with the densitometer, the sensitivity of various films can be compared (Fig. 8.8, right). However, the sensitivity of the complete film-screen system should be considered (see section 8.1.4).

Slope of film

The slope is derived from the size of the angle α (Fig. 8.6). The slope of the film is steep when α is larger; the slope is flat when α is smaller. The slope determines the exposure range.

Film with a steeper characteristic curve (gradation curve) produces higher contrast images (higher film contrast), i.e., it converts the quanta into a greater density difference (see section 9.3 for more information on contrast).

- The larger α is, the steeper the gradation curve (the steeper the slope) and the narrower the film's exposure range.

Film with a flatter characteristic curve produces poor-contrast images (lower film contrast), i.e., it converts the quanta into a smaller density difference.

- The smaller α is, the flatter the characteristic curve (the flatter the slope) and the larger the film's exposure range.

The exposure needed to reach the optical density (veil + 1) is determined based on the characteristic curve. Film A requires an exposure of 1.5 and film B 2.1 (Fig. 8.8, right).

As already described, the exposure gradations (of each degree of blackening) is such that the exposure doubles every other step (Fig. 8.7). The measurement in the example (Fig. 8.8) shows that film B needs a dose four times higher than film A to reach a density of 1.0 (above veil). Therefore, film B is only 1/4 as sensitive as film A. It can also be calculated as follows:

- log (exposure B of film B) − log (exposure B of film A) = log 0.6 = 3.98 ≈ 4; (logarithm, see optical density and appendix).

Since the intensifying screens are the film's primary source of exposure, the sensitivity of the complete film-screen system should always be taken into account.

Reciprocity law failure

The same mAs product, e.g., 500 mAs, can be achieved with 50 mA at 10 seconds exposure and with 1,000 mA at 0.5 seconds exposure. Although both cases produce an image with the same mAs, that does not necessarily mean they have the same optical density. The reason for this is the reciprocity law failure (see appendix).

The optical density of film that is exposed with intensifying screens is not directly proportional to the intensity (dose); however, the following applies:

$$D(OD) = I \cdot t^p$$

p is the Schwarzschild exponent. It is virtually 1, i.e., when similar mAs products and similarly high mAs are worked with, the reciprocity law failure (Schwarzschild effect) is barely noticeable. But it should be taken into account during measurements.

8.2.4 X-ray film processing

The extent of progress in X-ray film processing becomes clear when we consider that before developing machines were introduced, it took nearly 30 minutes from the time the film was immersed in the development tank until final rinse and drying.

Today, machines do the work of film processing, in dark rooms or inside daylight systems.

The development of the X-ray film completes the image generation process that began with an exposure on a radiography system. The following describes the steps of film development:

- The individual silver halide crystals in the emulsion layer of the film (photographic layer) that were exposed to radiation (quanta) and light emission from the intensifying screen first form a latent image of the imaged object.
- During development, only the exposed silver halide crystals are converted into the image-forming metallic silver.
- Intermediate rinsing removes the rest of the developer solution from the emulsion layer (gelatin layer).
- In the fixing bath, all the silver halide crystals are removed from the emulsion layer (washed out), leaving only the blackened silver halide crystals in the emulsion layer.
- To remove all the remaining developer chemicals, the fully developed film is cleaned in a water bath using running water. The finished X-ray image now contains only the silver halide crystals that are not sensitive to light and chemicals.
- The developed X-ray image is dried, i.e., the surface moisture is removed from both sides of the film using squeeze rollers and warm air.

Developers significantly reduced the labor involved in film development. They eliminate corrections during development and establish the best conditions for consistently high image quality.

8 Image Receptor Systems

Developers consist mainly of the following components:

- Roller transport system for the film
- Water system to wet the film and maintain temperature of baths
- Circulation system to mix the developer and fixing solutions as well as the regenerators
- System to regenerate the solutions
- Drying system and hardening of the film layers

The order of the processing steps remains the same, whether in developing film manually or by machine.

Manufacturers of developers provide detailed user manuals, including instructions on how to clean and maintain the systems and to dispose of chemicals and recover silver.

See chapter 9 for troubleshooting image quality. Consult the appendix for additional references on X-ray film development.

8.2.5 Intensifying screens

To review: since X-ray film is basically photographic film, its sensitivity to light is considerably higher than to X-rays, (approx. 95% to 97%). Only about 3% to 5% of the X-ray quanta is absorbed by the emulsion layer of the film.

That is why intensifying screens are used in radiography. Film-screen systems are the combination of film and intensifying screens.

The screens are constructed similar to film. Their light-sensitive layer consists of highly fluorescent phosphors that convert X-rays into light. When very short wavelength X-ray quanta strike the atoms in the phosphor crystals, longer wavelength UV or visible light waves are created. The process is known as luminescence.

Figure 8.9 shows the schematic structure of an intensifying screen: Protective layer (1) to prevent damage and electrostatic charge. Phosphor layer, fluorescent layer (2). Background or reflection layer (3), increases light yield,

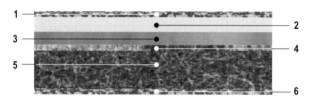

Figure 8.9 Structure of intensifying screen

146

but also unsharpness. Adhesive layer (4) to attach the phosphor layer. The base (5), usually made of polyester, thickness approx. 0.025 mm. Back layer (6).

In radiography, various phosphorous substances with different intensifying characteristics are used. There are two types of screens, calcium tungstate crystals ($CaWo_4$), or rare earth in use since 1973. Rare-earth elements are chemicals from the lanthanoid series: lanthanum (La, atomic no. 57), gadolinium (Gd, atomic no. 64), europium (Eu, atomic no. 63) or terbium (Tb, atomic no. 65).

Detective quantum efficiency (DQE)

Unabsorbed X-ray quanta do not really contribute to imaging. The detective quantum efficiency (DQE) is the percentage of X-ray quanta that are actually detected by the receptor. For intensifying screens, the DQE is a maximum of 30%. See section 8.6 for more information on DQE.

Some of the X-ray quanta impinging on the phosphor layer of the screen penetrate the layer; the rest are absorbed and emit the light to create the image.

The DQE, i.e., the absorption and conversion of X-ray quanta into visible light, is considerably higher in rare-earth screens than in $CaWo_4$ screens. To produce the same image quality, rare-earth screens need only about half the dose. Their DQE is better by a factor of 2. For this reason, $CaWo_4$ screens are barely used nowadays.

To manufacture intensifying screens, the approx. 2 µm to 10 µm phosphor crystals are added to a synthetic coating. This mixture is used to coat the base.

Intensifying screens with thicker phosphor layers of approx. 0.2 mm have a higher density of phosphor crystals (approx. 150 mg/cm^2). They absorb a larger percentage of the X-rays (quanta) and glow more brightly; they intensify the light that exposes the film more intensely than intensifying screens with thinner phosphor layers.

A thicker phosphor layer, however, does produce more scattering of both X-rays (quanta) and light emitted by the phosphor crystals than its thinner counterpart. As a result, thicker phosphor layers have greater screen unsharpness.

Screen unsharpness

Screen unsharpness is really the capability (limit) for the maximum edge detail or the finest resolution. Resolution of gray steps, i.e., the tiniest details of an image (spatial frequency) recognizable by the human eye (see section 9.3). To support the different requirements of X-ray technology, intensifying

8 Image Receptor Systems

screens with a variety of thicknesses, i.e., with different intensifying factors, are manufactured.

Sensitivity of film-screen system

Since the intensifying screens are the film's primary source of exposure, the sensitivity of the complete film-screen system should also be taken into account. The method for determining sensitivity is described in the following.

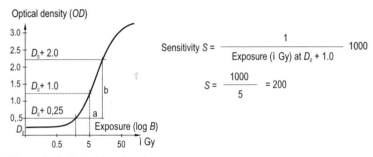

Figure 8.10 Sensitivity of a film-screen system

The measurement pictured in Figure 8.10 can be calculated from the measured values as follows: 1. Expose to normal radiation, develop under optimal conditions, and record density curve. 2. Measure the optical density of the veil D_0 and add 1.0. This example yields a required dose of 5 μGy. This would yield a sensitivity of 200, based on the definition.

Intensification factor, sensitivity class

Intensifying screens are characterized by their intensification factor and their screen unsharpness. The higher the intensification factor, the lower the image's resolution and the less dose is required. The lower the intensification factor, the higher the resolution (edge sharpness); however, the required dose is also higher.

Figure 8.11 lists the sensitivity classes in use today and their definitions. The required dose and the sensitivity class derived from it are absolute values. The intensification factor (also called the relative sensitivity factor) is a relative number always set to 1 based on high intensification rare earth screen (also called a universal screen) with a speed class (SC) = 200. The sensitivity or speed class SC of film-screen systems can be calculated using the following formula:

- $SC = (1/(D_0 + 1.0)) \cdot 1000 = 1000/5 = 200$

8.2 Film-screen Systems

Sensitivity class (SC)	Dose required (µGy)	Intensification factor	Resolving power (LP/mm)
50 (very high definition)	20	1/4	approx. 4.8 - 10.0
100 (high definition)	10	1/2	approx. 4.0 - 7.5
200 (universal screens)	5	1	approx. 3.4 - 5.5
400 (highly intensifying)	2.5	2	approx. 2.8 - 4.6
800 (very highly intensifying)	1.25	4	approx. 2.4 - 3.8

Figure 8.11 Intensification screens are divided into different speed classes that define the required dose and the resolution for X-ray film imaging

or similarly, for 2.5 µGy:

- $SC = (1/(D_0 + 1.0)) \cdot 1000 = 1000/2.5 = 400$, etc.

For clinical practice, intensifying screens are organized into the following speed classes. The radiographer needs to select the film-screen system that best suits the diagnostic situation.

Because they offer better protection against radiation, rare-earth screens are predominant. Figure 8.11 lists the relationship between the required dose (µGy) and the resolution for rare-earth screens. See the appendix for information on Gray (unit).

The different types of intensifying screens are identified according to their diagnostic application.

- Screens that produce the most detail (SC 50) are no longer permitted.
- The finely detailed rare-earth screen (SC 100; from 80 to 140) has a relatively low intensification but does produce images with the finest detail. These intensifying screens are allowed in specific cases only, e.g., images of the peripheral extremities for certain rheumatologic studies.
- The high-intensification rare-earth or universal screen (SC 200; from 160 to 280) represents a compromise between the effects of intensification and image sharpness. Application: peripheral extremities, shoulder, cervical spine, skull, chest (under certain conditions).
- The highest intensification rare-earth screen (SC 400; from 320 to 560) features the lowest resolution for the shortest exposure time. Application: extremities (upper thigh, hip joint), skull, thoracic and lumbar spine, pelvis, abdomen, gastrointestinal tract, chest.
- The ultra-high intensification rare-earth screen (SC 800; from 640 to 1,120) features the lowest resolution for the shortest exposure time. Application: lumbar spine (pregnancy), enlargement radiography, urinary tract.

8 Image Receptor Systems

Front and back screen

The front screen is closer to the object than the back screen. The back screen is often thicker, i.e., intensifies more than the front screen. Due to the absorption of the X-ray quanta in the front screen and the emulsion layer of the film, fewer X-ray quanta reach the back screen. If the back screen is thicker, however, it emits just as much light as the front screen. In this way, both emulsions of the X-ray film are equally exposed.

Intensifying screens with different thicknesses are marked accordingly and should not be inserted into the cassette in the wrong position. A higher screen unsharpness would result.

Crossover effect

The light of an intensifying screen can, under certain circumstances, penetrate the film emulsion and film base to expose the emulsion on the other side of the film base. This is termed crossover effect and causes unsharpness. The light that travels further to the emulsion on the other side is more scattered. Manufacturers can minimize this effect by modifying the structure of the film accordingly.

Care of screens

Intensifying screens should be checked regularly for dirt, scratches and other damage. The manufacturer's cleaning instructions should be followed.

Tube voltage and intensification

The absorption coefficient of the phosphors in the intensifying screens depends on the beam quality, that is, the tube voltage U_R (kV) and the filtering (section 3.3). Because the light generated by the intensifying screen is proportional to the amount of radiation absorbed, the intensification effect is somewhat dependent on the kV. This fact has to be taken into account during measurements and sensitivity comparisons. This relationship between kV and intensification (also called energy pathway of the screen) is accounted for in the exposure and point tables. See chapter 9 for the exposure and point tables.

Graduated screens

Graduated screens, also called plus/minus, comparative, gradient or differential screens, do not have the same intensification factor over their entire phosphor surface. Instead, the intensification is gradual. A plus sign "+" is used for the higher intensification side and a minus sign "–" for the lower intensification side.

Gradient screens are used for objects with great absorption differences. Typically, "+" is used for skull and "+-+" for extremities or the spinal column.

The varying intensification is used with the higher-intensification end ("+") on the side of the object that is more absorbent, that is, the thicker or denser end. The less absorbent end of the gradient screen ("–") is used on the end of the object that absorbs fewer X-ray quanta.

Instead of gradient screens, an additional absorption body can be laid on an object with highly varying thickness or density. Often, appropriately shaped filters are attached inside or onto the collimator.

Mammography screens

The need for fine detail is extremely high in mammography. For a long time, intensifying screens were avoided due to their screen unsharpness. The disadvantage was higher radiation dose.

Today, fine-detail rare-earth screens are usually used as the back screen, because the radiation scattered by the phosphor crystals does not reach the X-ray film and therefore cannot cause unsharpness. Dose is reduced and screen unsharpness kept to a tolerable level.

Digital imaging systems will eventually replace film-screen systems in mammography, as well.

Blue and green emitting screens

Until the advent of rare-earth screens, all intensifying screens had calcium tungstate ($CaWo_4$) crystals in the luminescent layer that emitted "blue" light. X-ray films were adjusted accordingly to be sensitive in the blue spectrum.

Because of their high specific gravity (higher atomic numbers, see appendix), rare-earth screens absorb a higher percentage of X-ray quanta than $CaWo_4$ screens. Rare-earth screens also generate more light per absorbed quanta; this light, however, is in the green spectrum.

The color spectrum consists of the pure, unmixed colors produced when light is physically broken down. The spectral ranges are based on various wavelengths (nm) (see chapter 2 for more on wavelengths).

- Violet   390 to 410 nm
- Indigo   410 to 430 nm
- Blue   430 to 500 nm
- Green   500 to 560 nm
- Yellow   560 to 600 nm
- Orange   600 to 650 nm
- Red   650 to 770 nm

8 Image Receptor Systems

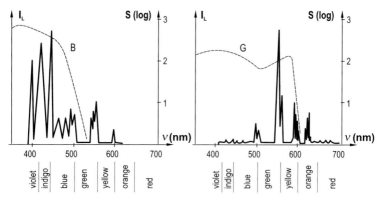

Figure 8.12
Spectral ranges: B) blue film-screen system. G) green film-screen system.

Available X-ray film is sensitive to either green or blue light to match the green- or blue-emitting intensifying screens (Fig. 8.12). To avoid premature exposure, the lighting in the dark room has to be adjusted to the X-ray film: green light for blue-sensitive film and red light for green-sensitive film.

Resolution

The resolution, i.e., the reproducibility properties of a multi-component system, is calculated from the modulation transfer function (MTF) of each component. (See chapter 9 for more on MTF).

The resolution of a film-screen system rests on both components: the resolution of the film and that of the intensifying screen.

8.2.6 Conventional film-free procedures

In the past, attempts were made to replace X-ray film with paper or synthetic screens. These "silverless" processes (e.g., xeroradiography or electroradiography) were not widely accepted. The main reason is probably that X-ray film had already reached very high standards of sensitivity, image quality and reliability. Nonetheless, a brief discussion of both procedures follows.

Xeroradiography

In place of X-ray film, xeroradiography uses an electrically charged selenium plate in a light-tight cassette. This layer is nonconductive when unexposed. During radiation, the conductivity increases pointwise proportional to the absorbed dose (quanta). This changes the distribution of the electrical charge on the surface. The charge distribution (of the latent image) is developed or made visible by dusting it with toner powder, similar to the way a copy machine works. The toner image is then covered with a synthetic coating. (See selenium drum, section 8.5 for information on selenium layer).

Electroradiography

In electroradiography, the imaging X-rays ionize a gas that fills a shallow gas chamber under high pressure. The ions produced by the radiation (quanta) are accelerated by an electrical field that distributes electrically charged particles on a screen. As in xeroradiography, this latent image is developed using toner.

8.3 Imaging Plate Systems

An alternative to film-screen imaging is digital luminescence radiography (DLR), which is performed using storage phosphors. This is also known as digital imaging plate radiography or computed radiography (CR).

Radiography using imaging plates continues to be used frequently. Further developments of new imaging plates with structured phosphor layers, for example, and special scanners for reading out the image information as well as the comparable image quality with flat detectors, particularly in thorax and skeleton exposures, continue to validate this alternative to flat detector systems.

Imaging plate image acquisition systems have several advantages over film-screen systems:

- Film savings: a reusable imaging plate replaces the single-use film.
- Dose savings with at least equivalent image quality.
- Major latitude of exposure due to a straight gradation curve and consequently almost no faulty exposures.
- Digital image processing and image editing.

Chapter 9 describes the use of analog-digital conversion to generate digital images from analog television signals and the diverse possibilities of digital technology for image processing. The method described here generates the digital image from the measured values supplied directly to the computer.

Figure 8.13 illustrates the basic structure of an imaging plate: layer for protecting against chemical and mechanical influences (1) and for preventing electrostatic charging, special semiconductor plate (2), e.g., barium-fluorohalide compound doped with europium (for information on emulsions, see section 8.1, see appendix for halides); layer base (3), e.g., polyester; X-rays (4).

8 Image Receptor Systems

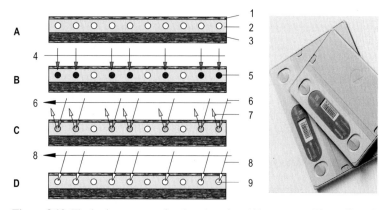

Figure 8.13 The basic structure of an imaging plate (A), exposure (B), reading (C), deleting (D), and cassettes with imaging plates

As a result of intentionally included traps (5) and corresponding doping between the conducting band and valence band of the semiconductor layer, the electrons excited by the X-rays (quanta) are captured by the available traps. Depending on the phosphor type, these traps remain stable for many hours, i.e., the imaging plate maintains a latent image after conclusion of the exposure.

If the imaging plate is scanned line-by-line by an infrared laser beam (6), every electron in the traps (or points of adhesion) that is hit by the laser beam emits a certain light quantity (7). The dose absorbed during exposure and the number of excited electrons are proportional over a large range (large dynamic range). The emitted light quantities (7) are intensified, digitized, supplied to an image processor, and stored after being converted to electrical impulses. The result images are available on the monitor as stored images.

Uniform, intense illumination (8) is used to delete the residual information (9) still contained in the imaging plate. The imaging plate is then available for another exposure.

X-ray systems that use film-screen cassettes for imaging can just as easily use cassettes with imaging plates. This means that current X-ray systems with cassettes and the customary exposure technique do not need to be changed. A cassette with an imaging plate is simply used instead of the film-screen cassette.

Due to the high dynamic range, one of the main advantages of the imaging plate is the relatively high insensitivity to faulty exposures, e.g., during bed exposures or exposures using a free exposure technique. They are used for spotfilm devices, intensive exposures and bed exposures or for standard radiography, such as for chest, gastrointestinal tract, urogenital system, breast, bone, and tissue.

Imaging plate

The image receptor of the imaging plate, i.e., the active "detector layer", has a design similar to intensifying screens. It is made of small, approx. 3 µm to 10 µm, phosphor grains distributed in a binding agent deposited on the support layer. The X-ray radiation (quanta) is converted in the phosphor layer of the imaging plates into light, as in the case of intensifying screens.

When the X-ray quanta impinge on this layer of the imaging plate, only a certain portion of the light is emitted immediately. Imaging plates are able to store the received information point-by-point as a latent image for a longer period of time. This storing capability is achieved in that additional traps are generated by including specific grid defects.

Reader, scanner

A laser beam uses the light of a certain wavelength (typically in the red to green spectral range) to scan the imaging plate. In this process, every point hit by this fine laser beam emits a light quantity which is proportional to the dose absorbed during the exposure. In this way, the image information is read out point-by-point and line-by-line. This readout process is used in readers/scanners on the basis of the flying-spot principle.

Scanners function on the basis of the line-scan principle with the same basic functions. In contrast to the flying-spot principle, the oscillating mirror which directs the laser beam line-by-line over the individual lines is not needed to read out the image information (Fig. 8.14 M). In line scanners, an entire line of pixels is read out at once. The laser light source consists of a series of laser diodes.

Fig. 8.14 shows the basic sequence of image creation using an imaging plate (CR): For light emission (photostimulation), i.e., reading out of the image information, the electrons stored in the traps of the detector or phosphor layer are excited by a laser beam (L).

The readout systems are referred to as readers or scanners. These are necessary for reading out the latent image information stored in the phosphor layer and for supplying it in digital form to a computer for evaluation (R).

After conversion into electrical signals and their intensification (Fig. 8.14-V), these are digitized in an analog-digital converter (A/D), stored (S), and supplied to a computer for image reconstruction (PC). A digital-analog converter (D/A) makes the image information available at the image processing station (M). The network connection or the transfer is performed from the local memory to a laser camera.

8 Image Receptor Systems

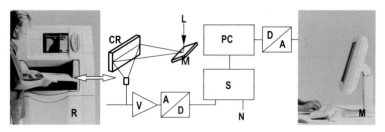

Figure 8.14
Schematic diagram of the process in a scanner or reader for imaging plates

The line-by-line laser beam scanning is also referred to as a scan operation. This scan operation entails a considerable amount of time. The storage phosphors have a natural afterglow time in the interaction with the laser light. If the scan operation were to be too fast, part of the information of the previous phosphor grain would be included in the current one. This would reduce the image sharpness. A slower laser speed adapted to the afterglow time prevents this effect.

Detective quantum efficiency DQE

The detective quantum efficiency DQE for imaging plates is between 30% and 50%. For information on DQE, see section 8.6.

Therefore, the absorption of the X-ray quanta depends on the chemical composition of the phosphor grains and on the thickness of the phosphor layer. As in the case of the screen unsharpness of the intensifying screens, a thicker layer limits the maximum definition or lowest resolution capacity for imaging plates (see chapter 9).

Optical density *OD*

Figure 8.15 shows that the latitude of exposure is significantly greater as a result of the straight course (B) of the gradation curve for imaging plates.

In the case of an exposure using a film-screen combination, an insufficient dose (log B in µGy) or insufficient X-ray quanta intensity results in an underexposure, i.e., in insufficient film blackening and a loss of diagnostic information due to the veil. In the case of film-screen systems, overexposure results in a loss of contrast due to the flattening gradation curve in the shoulder region (for information on veil and shoulder, see section 8.1).

The comparison of the two exposures in Figure 8.15 shows:

A1) Film-screen exposure with underexposure

B1) Imaging plate exposure with underexposure

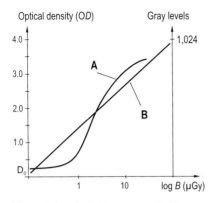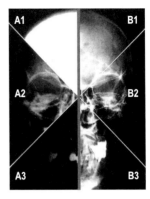

Figure 8.15 Gradation curves of a film-screen system (A) and an imaging plate system (B) and the comparison of an imaging plate exposure (B1-B3) with a film-screen exposure (A1-A3)

A2) Film-screen exposure with correct exposure

B2) Imaging plate exposure with correct exposure

A3) Film-screen exposure with overexposure

B3) Imaging plate exposure with overexposure

An image histogram is compensated to adapt the optical density to the limited density range of the viewing medium, e.g. hardcopy or monitor. A histogram is the graphical representation of a frequency distribution in the form of bars or curves corresponding to the frequency of the measured values.

After a histogram of the grayscale content is created for the original image, grayscale regions with significant information (pixels) are expanded, i.e., the contrast is increased. Grayscale regions containing less information (pixels)

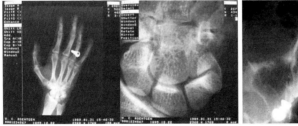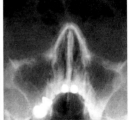

Figure 8.16 Example of the digital image processing possibilities for the same X-ray image (left). Bone and soft-tissue structures are visible. Certain density ranges can also be highlighted or suppressed via windowing (for information on windowing, see section 9.3). Paranasal sinus exposure (right).

are compressed, i.e., the contrast is decreased. If significant image information is contained in a limited grayscale region, the result is best when the region can be expanded via histogram compensation (Fig. 8.16).

8.4 Image Intensifier Television Systems

The pioneers of radiology sat in dark rooms and stared at a barium platinum cyanuric fluorescent screen to see the unsharp contours of the chest and heart (Fig. 8.17). In 1955, the first electron-optical image intensifiers were introduced. The large-surface, photoconductive X-ray tube was developed in 1956. The image intensifier television became established in the 1960s.

In contrast to the film-screen and imaging plate techniques, the image intensifier television technique (Fig. 8.18) has the advantage of direct monitor viewing of the result images from dynamic processes. Fluoroscopic scenes or single exposures can be recorded or stored at the same time and repeatedly called up from the storage medium and viewed.

X-ray systems with an image intensifier can be divided into two groups on the basis of their image receptor technology.

- Analog imaging:
  It seems outdated in light of modern digital image processing capabilities. However, it still has its place in radiology. The use of such systems depends on the diagnostic requirements and also on cost.
- Digital imaging:
  This seems to be the method of choice in view of modern and further developing digital image processing capabilities. However, other technol-

Figure 8.17 The Siemens KLINOSKOP X-ray system from 1907 was already equipped with a patented compression diaphragm. SIRESKOP undertable fluoroscopy system from 1953 with fluorescent screen and cassette (right).

8.4 Image Intensifier Television Systems

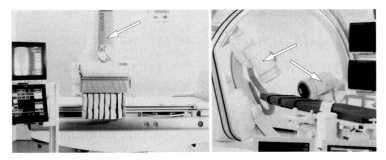

Figure 8.18 Modern X-ray systems with image intensifier: For an undertable fluoroscopy system with a 40-cm I.I. (left) and a 2-plane angiography system (right) with a 33-cm I.I. for each plane.

ogies such as flat detector systems present serious competition for today's solutions using image intensifiers.

Today, every image intensifier television system can be equipped with digital image generation and image processing components.

Figure 8.19 schematically shows an I.I.-TV imaging chain. Image intensifier output fluorescent screen (1) with a tandem lens system (2) for transmitting the I.I. result image to the signal plate of the TV tube or the CCD sensor (5).

The television camera (4), also referred to as the TV camera for short, is part of the television chain which includes the television pickup tube or the CCD sensor (5), the TV iris diaphragm (3), and electronic units, etc. The central TV control unit (6) and automatic dose rate control (7) are integrated in the generator.

The main components of an image intensifier chain are:

- The image intensifier tube,
- The television tube or a CCD sensor as well as
- A monitor (tube monitor CRT or flat screen LCD, see section 8.7).

Every photoelectron accelerated in the tube voltage potential of approx. 25 kV from the I.I. input screen to the I.I. output fluorescent screen generates approx. 1,000 light quanta on the I.I. output screen. The fluoroscopic image intensified several thousand times in this manner with respect to its brightness is scanned via a television tube or transmitted to a CCD sensor. The analog result image is transmitted via the central TV control unit or a digital system to a monitor.

Since the technology of digital imaging and image processing became available, 70 or 100 mm single-image cameras are no longer used in modern X-ray systems, as well as the 35 mm roll film cameras still sometimes used

8 Image Receptor Systems

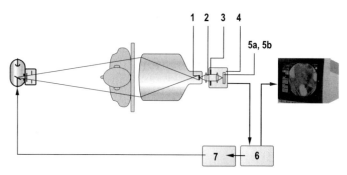

Figure 8.19 Basic structure of an I.I. TV chain for fluoroscopy with the TV tube or the CCD sensor (5)

in cardiac angiography. To create exposures with single-image or roll film cameras, a light channel distributor positioned between the I.I. output and the TV tube input screen is used to deflect the light quanta through a semitransparent mirror and related optics to the cameras (see section 8.3.3, Fig. 8.25). In the digital technique, the function of the roll film camera is assumed by a special control module (cine mode).

8.4.1 Image intensifiers

The electron-optical image intensifier (I.I.) is used to increase the brightness of X-ray images. It converts the radiation pattern, the impinging X-ray quanta, at the CsI I.I. input screen (as the scintillator) into an analog, brightness-enhanced light quanta image which is electronically directed to the I.I. output screen. The information lost at the I.I. input can no longer be included in the image reconstruction. This is fundamentally true for all image receptor systems, e.g., also for film-screen and imaging plate systems when information is absorbed by the cassette-holding frame.

However, the input window of the image intensifier (Fig. 8.20-2) must meet two requirements. These include: 1) the strength to withstand the pressure difference between the vacuum in the interior of the I.I. tube and the atmospheric external pressure, and 2) the highest possible permeability for the incident X-ray quanta. An external air pressure of approx. 1 kg per cm^2 is exerted on the I.I. input window of a 33 cm image intensifier. This is equivalent to the approximate weight of a mid-size passenger vehicle distributed over the entire surface. These opposing requirements are best met by a thin metal input window (typically 0.8 mm to 1.2 mm of aluminum). The transparency of these input windows is up to 95% for ICRU radiation quality.

Opposing requirements are also placed on the input fluorescent screen (Fig. 8.20-3). It is to absorb the greatest possible amount of X-ray quanta while

8.4 Image Intensifier Television Systems

not being too thick since as the layer thickness increases, the scatter of the quanta and consequently of the generated light within the phosphor layer also increase (see section 8.6). This effect also applies to all imaging systems (for information on screen-film unsharpness, see section 8.1).

Figure 8.20 (left) explains the creation of image information via X-ray quanta:

Quantum Q1: absorption in the CsI fluorescent screen (Fig. 8.20-3) provides the optimal contribution to visualization.

Quantum Q2: scatter in the input screen (2) and absorption in the CsI fluorescent screen (3); contributes to visualization. However, since it was scattered in the input screen, the image quality is not as good as for quantum 1.

Quantum Q3: scatter in the input screen (2), no absorption in the input fluorescent screen (3), no contribution to image information.

Quantum Q4: absorption in the input screen (2), no contribution to the image information.

Quantum Q5: no absorption in the input and fluorescent screen (2 and 3), no contribution to the image information.

Fig. 8.20 (right) illustrates the basic structure of an image intensifier: the image intensifier is made of a highly evacuated electron tube (5); its input screen (2) is covered by a cesium iodide phosphor layer CsI (3). The X-ray quanta are converted directly into fluorescent light (as in the case of intensifying screens and flat detectors with CsI scintillators). Incident X-rays (1). I.I. input window, approx. 0.8 to 1.2 mm of aluminum (2). Cesium iodide input fluorescent screen (CsI), conversion of X-ray quanta into light quanta (3). The electrons travel from the photocathode (4) into the interior of the I.I.

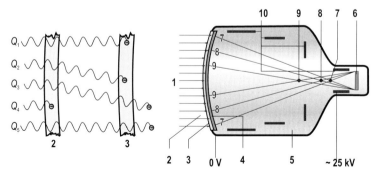

Figure 8.20 Absorption and scatter at the I.I. input (left), these effects also occur analogously in other image receptor systems. Basic structure of an image intensifier tube (right).

tube of the all-metal ceramic vacuum vessel (5). Output fluorescent screen (6), approx. 1% of the surface of the input fluorescent screen. Geometric point of intersection of the photoelectron paths (7) for a large I.I. input format as well as for enlarged visualizations (8 and 9), often also referred to as zoom. Electrodes of the electron optics (10) for deflecting the photoelectrons to the output fluorescent screen.

The radiation beam is attenuated differently when penetrating the fluoroscopy object. The radiation pattern resulting behind the patient in this process impinges on the input fluorescent screen (3) at the entrance of the image intensifier tube. The photocathode (4) coupled with the fluorescent screen on the backside generates light quanta (photons) in every pixel proportional to the number of incident X-ray quanta. The light quanta excite the photocathode to emit photons (photoelectrons).

In the case of today's customary low fluoroscopy dose rate, the CsI input screen (3) lights up relatively weakly. However, its light causes electrons to be emitted into the interior of the I.I. tube (5) as a result of the photoeffect from the adjacent photocathode (4). At points that are exposed more intensely, the number of photoelectrons is greater and vice versa. The two system layers, CsI input fluorescent screen and photocathode (2 and 3), therefore generate a photoelectron image from the X-ray image.

The photoelectrons are accelerated by a high potential difference of approx. 25 kV toward the output screen (anode) (comparable with the acceleration of electrons in the X-ray tube). The photoelectrons are focused electron-optically using the electrodes (10) generating electrical fields. They impinge on the photoelectron-sensitive viewing fluorescent screen (6, output screen of the image intensifier) and generate a very bright, smaller visualization of the X-ray fluorescent screen image (typically approx. 25 mm in diameter).

The energy supply in the electrical field of the image intensifier and the reduction of the image size result in the brightness enhancement of an image intensifier being approx. 20,000, i.e., the output screen (6) shows the image approx. 20,000 times brighter than the input fluorescent screen.

Electron-optical magnification

The conversion to a smaller image input format results in a magnified visualization of the recorded object at the output fluorescent screen. This is referred to as format changeover, magnification, zoom formats, or zoom levels (see chapter 9).

There are different I.I. input sizes with corresponding zoom levels to meet the diagnostic or interventional requirements (Fig. 8.21).

Converting the voltage at the electrodes of the focusing cup (Fig. 8.20-10) changes the image scale of the fluoroscopy object from the I.I. input screen

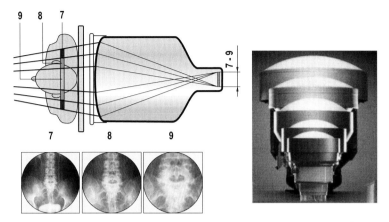

Figure 8.21 Left: Effect of magnification for different image intensifier input formats (zoom). Right: different image intensification sizes with input screens of 23 to 40 cm.

to the output screen. This magnification or size reduction entails a change of the conversion factor Gx in relation to the surfaces of the two input formats.

When working with the smaller input format, it is necessary to increase the dose rate by the amount of the aspect ratio. Therefore, the transition to a smaller format requires a change in collimation at the I.I. output screen using a TV iris diaphragm (see Fig. 8.19-3). When electronically switching to a smaller I.I. input format, the collimation of the useful beam in the primary collimator at the tube output is also adapted to the I.I. input format using an iris diaphragm (see Fig. 4.3).

Conversion factor

The quality of an image intensifier is assessed among other things on the basis of its intensification properties. These are described preferably by the conversion factor Gx. This factor corresponds to the ratio of the I.I. output signal (the luminous density of the output screen) to the input signal (the dose rate).

The intensification factor is the increase in luminous density ($cd \cdot m^{-2}$) by the factor corresponding to the aspect ratio between the input and output fluorescent screen (cd = candela, see appendix).

The conversion factor of an electron-optical image intensifier Gx has the unit: $cd \cdot m^{-2} / \mu Gy \cdot s^{-1}$. In relation to the dose rate irradiated at the I.I. input screen according to the ICRU radiation quality (in $\mu Gy \cdot s^{-1}$), this indicates which luminous density will be achieved at the output screen ($cd \cdot m^{-2}$). This refers to standardized ICRU radiation quality, i.e., 20 mm Al prefiltration and

7 mm Al first half-value layer thickness (see chapter 4). For information on ICRU and DIN, see the appendix.

In the case of image intensifiers with electron-optically switchable input fields (zoom), the luminous density also changes by the factor corresponding to the aspect ratio between the input and output screen.

Detective quantum efficiency (DQE)

The detective quantum efficiency DQE is up to 65% for image intensifiers. For information on DQE, see section 8.6.

TV iris diaphragm

For magnified images, the brightness decreases for an equally large illuminated output screen when switching to a smaller format (zoom) with a constant dose rate. However, the luminous density of the output screen of the image intensifier must always be adapted to the input sensitivity of the television chain.

The TV iris diaphragm is situated between the image intensifier output and the TV camera and is used to regulate the light at the I.I. output window. This is performed either by the dose rate control or additionally by a TV iris diaphragm (Fig. 8.19-3) which is electronically controlled to open more when switching to smaller input formats or to open less when switching to larger formats.

Electron-optical magnification also results in a more favorable modulation transfer function with improved detail recognition. The resolution, i.e., the imaging properties, of a system consisting of several components, is calculated from the modulation transfer function (MTF) of each individual component (for information on MTF, see chapter 9).

Vignetting

The properties of the optical and electron-optical components in image intensifiers are the reason that central image parts are displayed with slightly greater brightness than those on the edge. The decrease in the luminous density from the image center to the image edge can be attributed to the slightly greater image scale at the image edge. The vignetting (contour peripheral shadow) can be compensated for electronically.

Astigmatism, distortion

The focal distance of the lens is different for the points outside of the axis as a result of the planes running through the center axis of the imaging system (light waves have different angles of refraction). Object points are turned into lines in the image or straight lines become curved (concave or convex).

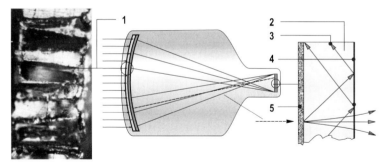

Figure 8.22 Image intensifier with CsI input screen and anti-reflection output screen with a glare trap

This effect (astigmatism, distortion) increases as the distance from the center of the image field increases, i.e., as the image intensifier increases.

Anti-reflection screen and glare trap

The aluminum input window which sustains an atmospheric load of 15 kN (1,500 kp) via a vaulted input surface of 1,500 cm^2, is only 1.2 mm thick. This window has a transparency of 93% (according to ICRU).

The section in Figure 8.22 shows part of the CsI input screen (left). The needle-shaped luminous crystals are perpendicular to the fluorescent screen plane and improve resolution as a result of their structure. The highly transparent input windows and the CsI fluorescent screen (1) achieve a particularly effective utilization of the X-ray quanta (see DQE).

High resolution and minimal structure noise are also required at the output screen (2) of the I.I. tubes. This is achieved by a very fine, highly compressed phosphor layer (5). The light impinging on the exit surface (4) at a certain angle no longer returns to the phosphor layer (arrows). It is absorbed by the specially prepared outer wall (3) of the glare trap.

8.4.2 Television pickup tube

In the case of X-ray television, the image content of the optical image resulting from the brightness differences is reduced in size on the I.I. output fluorescent screen. Via a lens (tandem lens system) this image is transmitted to the signal plate (Fig. 8.23-2), the target of the TV tube, is converted by this plate into electrical signals, and is relayed to the video amplifier (Fig. 8.23-1).

The optical image generates an equivalent charge image on the target (Fig. 8.23-2). The electron beam (3) scans this charge image line-by-line and dis-

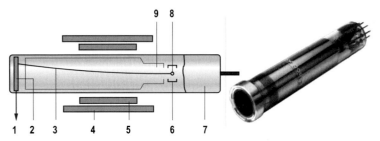

Figure 8.23 Basic structure of a TV tube: 1) electrical signals to the video amplifier; 2) signal plate (target); 3) the image scanning beam (scan) generated in the filament cathode; 4) focusing coil; 5) deflection coil; 6) Wehnelt electrode (for information on Wehnelt cylinders, see section 7.3); 7) vacuum glass vessel; 8) Filament cathode (for generating the electron beam); 9) Different grids for deflecting the electron beam. Right: 2/3" TV tube.

charges the individual positively charged pixels of the charge image. This generates a signal current which reproduces the optical image as an electronic image in the target. The electrical signals equivalent to the image are supplied to the video amplifier for amplification based on a specified standard. This is referred to as the video signal (1).

Scanning process

The image is scanned line-by-line by the TV camera and is reproduced on the monitor. The different scanning types are specified in European and international TV norms. CCIR norms apply to Central Europe (see appendix).

To prevent subjective image flicker on the monitor resulting from the optical capacity of the human eye, the refresh rate must be at least 50 Hz, i.e., 50 images (image scans) must be transmitted per second.

This scanning is performed using an electron beam that "scans" the image line-by-line from top left to right bottom, e.g., 50 times per second. Figure 8.24 shows the two variations of the image scan. There are two different scanning methods: interlaced scanning and progressive scanning.

Interlaced scanning (Fig. 8.24-A) is a cost-effective solution compared to progressive scanning (B).

In interlaced scanning (Fig. 8.24-A) a first half image (a) is scanned and then the second half image is scanned after grid return (c).

Two half images are scanned consecutively at a refresh rate of 25 images each per second. The scan duration for one half image is 20 ms. The joining of the two half images is perceived by the human eye at a refresh rate of 50 images per second.

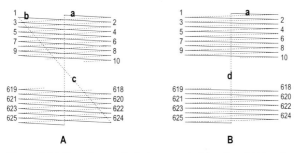

Figure 8.24 Scanning methods (scan modes) interlaced (A) and progressive (B) for a 625-line example

In the case of progressive scanning (Fig. 8.24-B) for single images and series, every line (a) is scanned without interlacing at a refresh rate of 50 images per second. After the lines are run through, grid return (d) is performed to scan the next image.

Lines, image repetition, line frequency

The CCIR standard requires image scanning at 625 or 1,249 lines for countries with a line frequency of 50 Hz (Central Europe), and 525 or 1,023 lines for countries with a line frequency of 60 Hz (e.g., the USA). The product of lines multiplied by the time into which an image transmitted is broken down is almost equal for the two standards : 625×25 (50 Hz) / 525×30 (60 Hz) or $1,249 \times 25$ (50 Hz) / $1,023 \times 30$ (60 Hz).

From a technical standpoint, the line frequency (50 Hz or 60 Hz) does not influence the refresh rate of 50 or 60 images per second.

In the same manner that the image is scanned line-by-line by the TV camera, the TV image is also reproduced from a number of lines that follow closely on each other.

In the case of digital imaging systems, a frame buffer is used to control transmission so that an optimum flicker-free image is displayed on the tube monitor at a refresh rate of 76 Hz.

Bandwidth

The geometric resolution is limited in the vertical direction by the line number. Lines are to be understood as a series of pixels which must be resolved due to the rapid scanning, comparable to the reading out of the imaging plate (transmission of spatial frequency, geometric resolution, see chapter 9).

The video signal of the TV tube is amplified in the video amplifier for transmission to and display on the monitor. The maximum frequency range of the

8 Image Receptor Systems

video signal, i.e., the maximum transmission range of the spatial frequencies, is referred to as the video bandwidth and is provided in MHz.

Signal-to-noise ratio

A dB value is provided together with the bandwidth. The dB (= decibel) (see appendix) is a dimensionless value which is provided, for example, for signal-to-noise ratio values. The signal-to-noise ratio is the ratio of the signal power to undesired interference power when transmitting the video signal (noise, noise ratio, signal-to-noise ratio). These signals are noise signals that are superimposed on the video signal due to the electronic image transmission systems and have a negative effect.

Positive values signify amplification, negative values signify damping. A typical value of -4 dB, for example, means a low noise ratio, signal-to-noise ratio. dB information is also known as the volume level in electronic amplifiers, for example, and indicates the input-to-output voltage ratio.

The noise ratio (signal-to-noise ratio) is sometimes also provided as a numeric ratio of the signal power to the undesired interference power, e.g. 1:2,500, which equals approx. -4 dB.

8.4.3 Image intensifier indirect exposure technique

Digitization advancements in radiography have rendered various radiographic techniques obsolete. Radiographic systems using sheet and/or roll film cameras are no longer manufactured and sold today. However, since they are still used in older radiography systems, a brief explanation is provided here.

Sheet film camera

Photography with 100 mm or 70 mm sheet film cameras (Fig. 8.25-3) was the preferred method for indirect exposure techniques in serial radiography from fluoroscopy. The cameras were able to record up to 6 images per second. The advantage was that the quality of the high-resolution image intensifier had the same image information content as that of direct exposure via film-screen systems. Film cost savings were achieved and the dose (typically 0.66 µGy) was reduced to approx. 30% of a full-size X-ray radiograph.

Sheet film cameras were able to accommodate up to 120 single films. After every exposure, the film was automatically changed. The exposed films were automatically developed in a developing machine.

The light distributor (Fig. 8.25-1) receives the image signal from the I.I. output screen (D) and directs it via the basic lens (D) to the TV camera lens (C). The image signal is deflected at a semi-transparent mirror (B) to the I.I. camera lens (multi-format or roll film camera).

8.5 CCD Systems

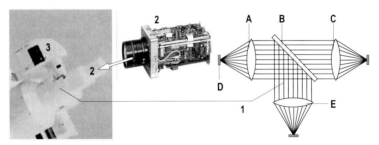

Figure 8.25 Using 2 or 3-channel light distributors (1), a sheet and/or roll film camera (3) is positioned next to the TV camera (2) for image generation so that additional X-ray films are also exposed. Light distributors with high-grade and strong-light tandem lens systems (1) distribute the light at approx. 10% to the TV camera (e.g., C) and 90% to the film cameras (e.g., E).

Roll film camera

The bright dynamic image of the I.I. output screen was also able to be filmed using kinescopes for function studies of quickly moving organs and processes. 35 mm kinescopes (roll film cameras) with an exposure rate of up to 60 images per second were used in angiocardiography.

Digitization

Digital exposure technology has replaced this method. Single exposures or imaging modes of 15, 30 or up to 60 images per second, with a frame buffer capacity of up to 50,000 images, are possible today. The television pickup tubes are also replaced by semiconductor image sensors, known as CCD sensors, to digitize the analog image information signals. For information on this and how analog image information signals are digitized at the I.I. output screen, see chapter 9.

8.5 CCD Systems

In radiodiagnostics digital fluorescent screen radiography uses CCD sensors (charge-coupled device) just like TV pickup tubes to convert the light of an I.I. output image into an electrical signal.

Long before its use in radiological diagnostics, CCD sensor technology (Fig. 8.26) was an established part of entertainment technology as image recording systems (e.g., camcorder). CCD systems and flat detector systems for digital real-time imaging are now an established part of medical imaging.

In general, the image information of the X-rays (quanta) in CCD systems are converted in a scintillator layer into light and directed to the CCD sensors.

8 Image Receptor Systems

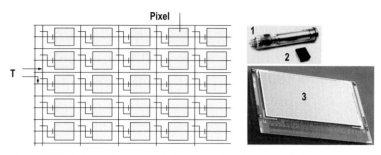

Figure 8.26 Basic CCD sensor and size comparison between a 2/3" TV tube (1) and a CCD sensor with a 0.5^2 k matrix (2). CCD sensors starting at a 1^2 k matrix are used in high-resolution TV chains. The picture on the bottom right (3) shows a CCD detector used in mammography.

Since the X-ray quanta were already converted into light at the I.I. output screen, it is transmitted directly to the light-sensitive elements which are attached to the semiconductor crystals of the CCD sensors in the form of a matrix. The charge image of the photons is converted into an electrical charge image in the semiconductor crystals of amorphous silicon (a-Si).

A CCD system is comprised of a number of silicon image sensors arranged in a surface-like manner (photocells = pixels). They string together to form a line. Line by line they yield a surface referred to as a matrix (for information on matrices, see chapter 9). The coupling to the light (the image of the image intensifier output screen in this instance) can be achieved via mirrors, lenses or optical waveguides.

They store the direct image of the I.I. output screen as an electrical charge image (see target for the TV tube), i.e., a direct visualization of the image brightness is produced in these photoelements, photosensors (pixels), as an electrical charge image.

After completion of an integration cycle, this charge image is read out via a transport register (Fig. 8.26-T) as an electrical signal, i.e., they are transported via a charge-coupled device pixel-by-pixel and line-by-line to the output (T) of the circuit.

After each individual pixel information has been read out, converted into a voltage value, accordingly intensified and stored in a special image repeat memory, the result image is available on the monitor.

In contrast to line-by-line scanning in the TV tube (only in the vertical direction), all stored charge packets (image information) are run through from one pixel to another in the transport register (T) (in the vertical and horizontal direction) in the case of CCD sensors. The pixels are printed in their horizontal and vertical arrangement in a matrix. The process for generating the image information is comparable to that of flat detectors (see section 8.6).

In addition to compact construction and low energy consumption, the advantages of the semiconductor technology with CCD sensors include:

- Virtually no aging (long service life)
- No microphony (conversion of mechanical into electrical oscillations that are superimposed on the video signal in TV);
- No inertness (the charge image is read out as an electrical signal);
- No geometric distortion;
- No burning-in.

Even if the image quality features are comparable in some respects to those of flat detectors, there are still significant differences, e.g., the detective quantum efficiency (DQE) and the dynamic range. For information on DQE, see section 8.7. The advances in flat detector technology in connection with digital image reconstruction and processing present serious competition for the image intensifier technology with CCD technology.

8.6 Flat Detector Systems

Flat detectors (FD), often referred to as flat panel detectors, are used today in radiographic systems for chest and skeleton, in cardiac angiography and increasingly in radiographic systems for universal and special fluoroscopy, universal angiography and mammography for radiographic imaging (Fig. 8.27). They are in the process of successively replacing radiographic technologies using film-screen and imaging plate systems as well as systems based on CCD sensors. Dynamic flat detectors capable of processing 30 images per second make their extensive use in diagnostic radiography possible. This development is ongoing.

The detectors used in imaging systems for medical radiodiagnostics are tiny, individual components. The layers of a detector element for generating

Figure 8.27 Flat detector PIXIUM^{TM} 4600 used in today's radiography systems (center), a mobile flat detector (left) during insertion into the bucky tray of a radiography system, and wall-mounted vertical bucky stand (right) with the flat detector

image information are approx. $0.5 \cdot 10^{-3}$ mm thick with a support layer of glass substrate of 1 mm to 2 mm.

The geometric dimensions of a detector are approx. $0.15 \cdot 10^{-3}$ mm. Strung together, they form line-by-line a surface referred to as a matrix and similar to the CCD systems described above. For pixel and matrix information, see chapter 9. The surface-forming sum of all detectors, including the electrical components and microelectronics, forms the system of flat detectors.

Image information creation

Scintillation is the creation of light (light flashes) when high-energy particles or electromagnetic radiation, X-rays in this instance, impinge on fluorescent materials (the scintillator). The interaction of X-ray quanta with the shell electrons of the atoms in the scintillator layer (cesium iodide crystals, CsI, or sintered ceramics, such as gadolinium oxysulfide, Gd_2O_2S) results in the photoeffect.

Figures 8.28 and 8.29 show the creation process of the image signals in flat detectors. The photoeffect, a general term for the photoelectric effect, occurs when X-ray quanta (1) are absorbed in the electron shells (K, L, etc.) of the atoms of the scintillator. They disappear completely. In this process, free electrons, known as photoelectrons, are emitted from these shells which can further ionize in the scintillator layer (6). The emitted photoelectrons (3) from the interior of the scintillator layer penetrate a semiconductor material (8), the photodiode (of amorphous, i.e., glass-like, silicon a-Si in this instance), and form an electrical charge (+, −) there. Photoelectrons scattered upward are reflected by the reflector (5) back in the direction of the photodiode.

The excellent semiconductor properties of the a-Si are used to produce the smallest possible photodiodes (Figures 8.28-8 and 8.29-8) with switching

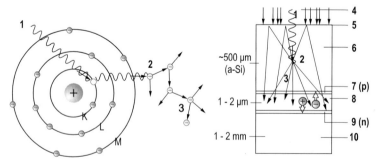

Figure 8.28 Schematic diagram of the photoeffect and photoemission at a detector element: An absorbed X-ray quantum (1) with an average energy of 53 keV (at a tube voltage U_R of 53 kV) generates a signal (2) of approx. 800 to 1,000 electrons (3).

8.6 Flat Detector Systems

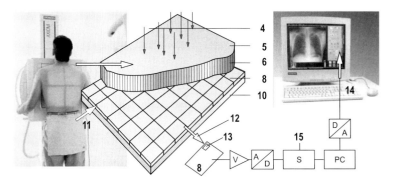

Figure 8.29 Principle of image information recording using a stationary detector on the basis of a-Si and the virtually immediate visualization of the result image on the monitor (14)

elements (Fig. 8.29-13), known as TFD switching diodes (thin film diode) or TFT switching transistors (thin film transistor).

Semiconductors are solid bodies that do not or only minimally conduct electrical current at low temperatures. The specific resistance of semiconductors (the electrical conductivity) depends on the foreign atom doping, the temperature, or the incident light. The basic element of most semiconductors is the positive-negative junction (pn junction). In this context, p (Fig. 8.28-7) stands for positive (conducting band) and n (Fig. 8.28-9) for negative (valence band, bonding band).

If the pn junction of a photodiode is subjected to light (via photoelectrons, light quanta), it becomes electrically conductive. The desired conductivity reduces the specific resistance of the semiconductor. The photodiode converts the light into an electrical charge which is read out via activated switching transistors (Fig. 8.29-13).

In contrast to the diode (photodiode, TFD), the transistor (TFT) has three consecutive zones with different conductivity modes, virtually a dual diode. As a result, two adjacent barrier layers, which are npn or pnp transistors depending on the conductivity mode, are obtained in each case.

Figures 8.28 and 8.29 show that the impinging X-ray quanta (4) are converted in the scintillator layer (6) into photoelectrons in a manner similar to the process at the input screen of image intensifiers or film-screen and imaging plate systems. The existing charge image of the photoelectrons in every semiconductor element (pixel) is converted into an electrical charge (charge image) in the semiconductor elements (8) deposited on a carrier material, i.e., the glass substrate (10). The photodiode covers approx. 60-70% of the actual surface of a pixel of the flat detector.

8 Image Receptor Systems

Figure 8.29 shows that every pixel is connected in the line direction (line driver, 11) with control electronics and in the column direction with an amplifier and analog-digital converter (A/D). The charges of the individual pixels flow along the column lines (12). The individual pixel charges are read out by parallel control of the switching transistors (13) within one line, then the next, etc., until the entire charge image has been read out. After conversion into electrical signals and their amplification, these are digitized in an analog-digital converter (A/D), stored (S), and supplied to a computer for image reconstruction (PC). A digital-analog converter (D/A) makes the image information immediately available at the image processing station (14). The network connection or the transfer is performed from the local memory to a laser camera (15).

For high image rates as in angiography or fluoroscopy, readout electronics with high bandwidths are available, i.e., the pixels must be resolved by quick readout (transmission of the spatial frequency).

Detective quantum efficiency (DQE)

In the case of a-Si flat detectors, the detective quantum efficiency DQE is up to approx. 80% as a function of the tube voltage U_R in a working range of 45 to 150 kV. For information on DQE, see section 8.7.

Dynamic range (optical density)

The latitude of exposure is significantly greater as a result of the primarily straight course of the gradation curve for flat detectors.

The needle structure of the CsI crystal (CsI) acts as an optical waveguide which conducts the light emitted in the green range of the spectrum with

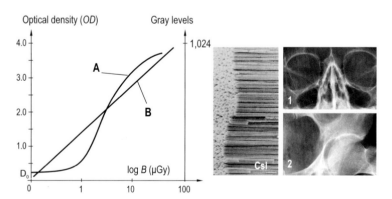

Figure 8.30 Compared to the curve of the optical density in film-screen systems (A), the advantage of flat detector systems (B) is obvious, compared to the properties of imaging plate systems

very little scatter onto the photodiode. The photodiode is also designed for high quantum efficiency in the green range of the spectrum (see "blue and green emitting screens" in chapter 8).

The large dynamic range, the film gradation curve, corresponds to a very large optical density range, also the density latitude. The gradation curve (Fig. 8.30, left) shows the relationship between the relative exposure log B (= exposure strength times exposure time, intensity in µGy in decadic logarithm scale) and the resulting possible dynamic range (optical density range, density latitude, density range). The dynamic range interprets the image properties (the quality) of the detector system on the basis of the density curve. It is defined as a decadic logarithm.

The large dynamic range of flat detectors allows, in approximately the same manner as imaging plate systems, a high tolerance with respect to erroneously input or switched generator exposure data.

Digital selenium technology

A special method based on selenium technology was introduced at the start of the 1990s particularly for large-format exposures, such as thorax radiography. Selenium detectors convert the X-ray quanta directly into electrical signals, not by photoemission as in a scintillator.

A layer of amorphous selenium (a-Se) is deposited on a cylinder surface, known as the selenium drum (Fig. 8.31), as the semiconductor image receptor. The surface of the selenium layer is electrically charged (L) prior to generating the X-ray image. During the interaction of the X-ray quanta with the selenium, electrical charges are released which reduce the surface charge of the selenium layer according to the radiation pattern. The electrical charge pattern corresponds to the X-ray image (B).

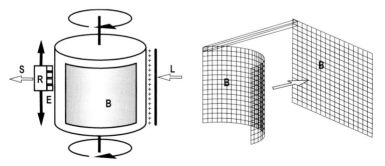

Figure 8.31 Principle of imaging on a selenium drum. Distortions in the segmental image format are subsequently corrected digitally.

8 Image Receptor Systems

Figure 8.31 shows that for reading out (R) the electrical charges, the drum rotates past laterally arranged measuring probes. The data is scanned point-by-point (pixel) and line-by-line until the entire charge image is read out. The image data (S) is available immediately, following amplification in special circuits or the analog-digital converter, as the result image (S) on a monitor or in a digital frame buffer for image processing. The detective quantum efficiency DQE is over 60% (for 60 kV tube voltage) and the dynamic range is comparable to that of flat detector systems. For information on DQE, see section 8.7. The selenium technology is a "direct digital" flat detector technology.

Flat detector technology: indirect/direct

The process of image generation in which image information is generated by converting the X-ray quanta directly into electrical voltage values is referred to as "direct digital" flat detector technology. This flat detector absorbs X-rays via a semiconductor based on amorphous selenium (a-Se), and converts them directly into electrical signals. Every pixel consists in this case of an electrode (TFD) and a switching transistor (TFT). Direct digital flat detector technology is used in the case of the selenium drum or also, e.g., in mammography systems.

Image generation with amorphous silicon (a-Si) is classified as "indirect digital" flat detector technology. The term indirect is used because the X-ray quanta are converted in a first step for image reconstruction into light (photoelectrons), which then generates the necessary electrical image signals in the photodiodes in a second step.

Detectors for computer tomography

Essentially two detector types are significant in CT:

- Ionization chambers usually filled with the noble gas xenon under high pressure, and
- Scintillator detectors in the form of crystals, such as cesium iodide (CsI) or ceramic materials, such as gadolinium oxysulfide (Gd_2O_2S).

The CT scanners already established today have one or multi-line detector arrays developed especially for these system types and segmentally arranged in the radius of the cone of radiation. Scintillators are primarily used today. In most cases, these are sinter ceramics such as gadolinium oxysulfide (Gd_2O_2S). Scintillator detectors of gadolinium oxysulfide have above all a very quick decay behavior of the light signal, i.e., they are better suited than CsI scintillators for quick dynamic processes. They have already replaced the cesium iodide and xenon scintillator detectors in CT. For information on computer tomography, see the references.

8.7 Summary of image receptor systems

Constant developments in the areas of chemistry, physics, mechanical engineering, electrical engineering, and electronic engineering provided the prerequisites for progress in medical radiodiagnostics.

- Film chemistry was a prerequisite for the successful film-screen systems.
- Vacuum electron tube technology provided the foundation for image intensifiers.
- Electrical engineering and electronic engineering provided the basis for multipulse and converter generators.
- The advances in electronics, microelectronics and processor technology paved the way for digitization.
- Semiconductor technology provided the basis for the development of flat detectors.

Fig. 8.32-1: In film (1C) the silver halides exposed by the intensifying screen (1A) to light (1B) visualize the image information. The non-exposed silver

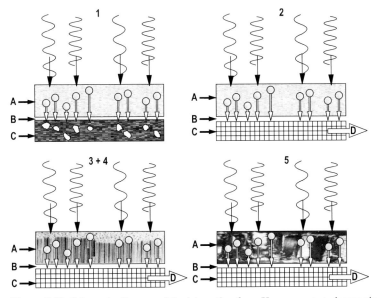

Figure 8.32 Schematic diagram of the interaction from X-ray quanta to image signals. A: absorption medium, and B: interaction medium. After conversion in the measuring medium (C) into electrical signals (D) and their amplification, these are digitized in an analog-digital converter, stored and supplied to a computer for image reconstruction. A digital-analog converter makes the image information immediately available at the image processing station.

halides are washed away during development. The X-ray image is then available.

Fig. 8.32-2: In the case of imaging plates (2A), the stored light parts form the image information which is emitted via a laser beam (2B) at a certain wavelength and converted for image reconstruction into electrical signals (2C).

Figures 8.32-3 and 8.32-4: CCD systems and flat detectors with indirect conversion ("indirect digital" flat detector technology) have scintillators of cesium iodide (CsI) or gadolinium oxysulfide (Gd_2O_2S) as the absorption medium (A3, A4). The X-ray quanta are converted into photons, i.e., into light as the interaction medium (B3, B4). This also applies in the case of image intensifier technology using CsI input screens. In the case of CCD systems in connection with an image intensifier, the absorption medium is the CsI input screen of the I.I. The light signals are converted in the CCD chip (3C) or in the photodiode matrix (a-Si) into electrical image signals.

Fig. 8.32-5: Flat detectors with direct conversion ("direct digital" flat detector technology) have amorphous selenium (a-Se) as the absorption medium (A5) (see selenium drum). The X-ray quanta are converted as an interaction (B5) in a capacitor matrix (C5) directly into electrical signals.

Even though flat detector technology is currently on its way to replacing all other imaging systems, it is still possible for new imaging system technologies to emerge as a result of ongoing developments.

The technology of the flat detector is not unlike the physical conversion of X-ray quanta into an image. X-ray quanta generate light in an intensifying screen, an imaging plate, or a scintillator (photons), which becomes the desired image information in the particular measuring medium (X-ray film, imaging plate, CCD sensor, flat detector).

Detective quantum efficiency (DQE)

Figure 8.33 shows a comparison of the detective quantum efficiency (DQE) of different absorber media. It must be noted that the represented values

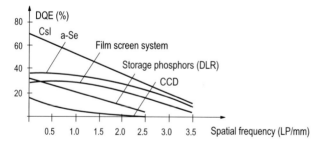

Figure 8.33 Comparison of the detective quantum efficiency (DQE) and achievable spatial frequencies (resolution) in different imaging systems

depend on different parameters such as exposure voltage, equivalent dose or prefiltration as well as the pixel sizes of the matrices, the absorber media (scintillator) or the optical coupling in I.I. TV systems with CCD sensor. For more information on DQE, see the appendix.

8.8 Image viewing systems

Different types of viewing systems (film illuminators) are available for evaluating or diagnosing translucent images (X-ray images). Image viewing systems for diagnosing translucent images are, like monitors, a basic part of every X-ray examination room or of every radiological department in private practices and hospitals.

Exposures in connection with analog or digital I.I. TV systems or solid-state detectors are visualized and evaluated directly on monitors. However, they can also be produced as film frames via laser cameras.

Result images from radiographic systems using digital image processing or the ability to digitize X-ray images can also be viewed or diagnosed and postprocessed on workstations in connection with a hospital network for the purpose of archiving and communication (see chapter 10).

8.8.1 Film viewing

Film viewing systems are wall-mounted, table-mounted, mobile or permanently installed electrically controlled film viewers for the viewing of X-ray

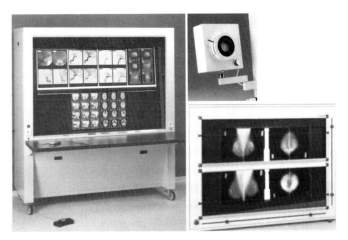

Figure 8.34 Film viewing systems: mobile film rack lifting system (left), wall-mounted film viewer and magnifying glass (right).

images (Fig. 8.34). A strong light source backlights a ground-glass plate with largely diffused light.

To adapt to the optical density of the X-ray images, the brightness of the viewing system is typically 2,000 cd/m^2. The size of the viewing surface must allow comparison of at least two exposures of the largest available format.

The image viewing surface must be delimited (darkening) on four sides, and zooming in to two to four-times the magnification must be possible.

8.8.2 Monitors, displays

The monitor must display the image with optimum quality, i.e., without distortion, with sharpness and with the correct grayscales. This is the last element in the imaging chain.

There are two monitor technologies:

- CRT monitor: CRT = cathode ray tube.
  Conventional monitor with a Braun tube and electron beam deflection, comparable with television and PC monitor tubes.
- LCD monitor: LCD = liquid crystal display.
  Flat panel monitor in which the screen is made of liquid crystals. This technology was already established prior to use in medical radiology. In German-speaking regions, the term display is used to express the difference to the CRT monitor.

In the case of CRT monitors, the analog video signal of the television camera is transferred via the central TV control unit as a live image or live scene to the monitor (Fig. 8.35). In the case of digital imaging systems, the conversion into analog image signals (digital-analog conversion) is performed prior

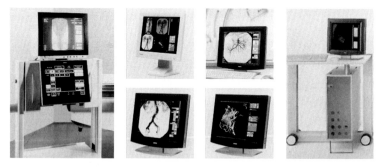

Figure 8.35 Mobile CRT monitor in the control room (left) and mobile image processing station with an LCD monitor (right). Center: different LCD black-and-white and color monitors.

8.8 Image viewing systems

to viewing on the monitor and the signals are also viewed as live images or live scenes. Digital-analog conversion is still used today for new LCD monitors since X-ray examination rooms without digital imaging may be equipped with LCD monitors.

CRT monitors transmit different line numbers per image reconstruction in accordance with the TV standard: 625/525 lines are standard (rare today) and monitors with at least 1,294/1,023 lines are referred to as high-resolution monitors.

Absolutely flicker-free images are provided in principle for every optical capacity at a refresh rate of approx. 76 Hz, i.e., 76 images are reconstructed per second on the monitor.

LCD monitors are becoming increasingly more common due to their now high-grade technology, as well as for a number of other reasons. As a result, the refresh rate of 76 Hz required for CRT monitors is the slower response time for LCD monitors, i.e., the response time corresponds approximately to the sensitivity of the human eye. No flickering is perceived. Advantages of LCD monitors are:

- Minimal weight
- Minimal space requirement
- Distortion-free image reproduction
- No irradiation
- No magnetic fields.

In the case of special examinations in the examination room, the viewing angle of the LCD monitors is not always sufficient. However, the horizontal viewing range is now approx. 170 degrees.

During examinations in which the absorption rates of the patient change, the monitor brightness is maintained via automatic signal control.

The contrast and brightness of the monitor image can be individually controlled via remote control. Since the quality of the monitor image is influenced by the room light and immediate environmental light, the monitor brightness can be adjusted via automatic regulation to the room brightness.

The location of the monitors is selected by the user based on specific requirements. In the examination room it is possible to use a mobile monitor cart with one or two monitors or to position one to eight monitors on ceiling-mounted monitor or display support systems (Fig. 8.36, right). The monitors are usually on the work tables in the control room (Fig. 8.36, left).

Monitors have different screen sizes and are identified by the diagonal size of their image tube, usually 44 cm and 54 cm. The screen size of LCD monitors is usually indicated in inches, e.g., 18" (46 cm).

8 Image Receptor Systems

Image processing and evaluation stations (workstations)

All radiographic systems equipped with digital imaging have a PC-based imaging system. The bi-directional data communication is performed within a PACS network via DICOM standard functions with connection to a RIS and HIS (see chapter 10).

Workstations (Fig. 8.36, left) are optimized for radiological evaluation and diagnosis. They are central communication components for patient and image data information management, frame buffering and image archiving.

 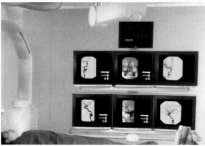

Figure 8.36 Digital image processing and diagnosing station (workstation) in a radiological department (left) and display support system with seven LCD monitors in the cardiac catheterization room (right)

9 X-ray Imaging Technology

X-rays, the basis for diagnostic radiography, are the information carriers for creating a picture of the inside of the human body.

Creation of the X-ray image

X-ray images represent important details for diagnosis and enable the selection of therapies (Fig. 9.1).

Independent of the X-ray system, nothing about this basic imaging position has changed since the discovery of X-rays (Fig. 9.1, left). Generation of an X-ray image (right): The different attenuation coefficients (density and thickness of the material to be irradiated) of bone, tissue, etc., produce different intensities on the image receptor.

Despite the rapid dissemination of digital imaging techniques, direct imaging on X-ray film remains the most used method of documentation world-wide. Conventional imaging techniques using, for example, 70 mm or 100 mm sheet film, sheet film exchangers or 35 mm cine-film (such as roll film for dynamic cardiac imaging) have lost importance due to digital imaging techniques. They are no longer offered by manufacturers.

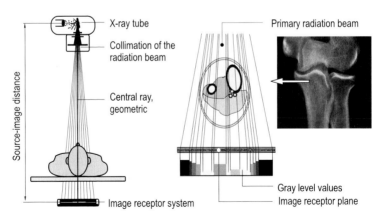

Figure 9.1
An X-ray projects parts of the body, the object, onto an image receptor system

The different X-ray systems used in radiology today with their different modalities for generating images are all subject to the laws of projection, imaging geometry and generator imaging data of radiological imaging. The following criteria must be observed to create a good X-ray image:

- The object must be positioned correctly taking into account anatomical factors.
- To obtain the right imaging properties such as contrast and sharpness, the exposures must be taken with the right radiological settings (kV, mAs) for the relevant object parts, their density and thickness.
- Taking into account the required dose and its reduction, a series of parameters can be used to analyze errors that produce inferior image quality (see 9.3).
- It is difficult or impossible to distinguish individual organs and organ parts that only have slight differences in tissue density. Contrast agents help accentuate slight differences in density (see section 9.3).
- To create an image of the diagnostically relevant area, the projection of the X-ray beam must be adapted to the placement and position of the diagnosed object. Independent of the technology of the image receptor system, the laws of projection and imaging geometry are basically the same for all radiological systems (see section 9.4).
- In digital radiography, there are numerous new imaging and processing parameters that have become indispensable (see section 9.2).

Radiographic references

To make a diagnosis, users must understand the projection angles and geometry of the different X ray systems. These are described in detail by the manufacturers of the respective X-ray system in the user manuals and data sheets.

Manufacturers also offer imaging references, exposure guides or setting instructions for the various X-ray systems with their different modalities for generating images. These documents describe in detail the correct positioning and placement of the patient for nearly every object to be imaged.

In addition to such manufacturer documents, the professional literature offers comprehensive and detailed descriptions of radiological settings (see 9.4.3) and references.

Uninterruptible power supply (UPS)

To circumvent power failures while using the imaging systems, an uninterruptible power supply independent of the public supply must be ensured for a certain period. Manufacturers offer uninterruptible power supplies (UPS) for emergency situations.

An emergency power supply automatically assumes the supply of power for 10 minutes when there is a failure of the primary power grid (for the entire system or subsystems depending on the configuration). This ensures continuous operation for a defined period or, at least, until the conclusion of an examination series.

9.1 Radiographic Parameters

The overall X-ray beam is a mixture of the X-ray spectrum generated at each moment of imaging (see chapter 2, X-ray spectrum). The setting of, for example, 100 kV yields different beam qualities in generators with different technologies (2, 6, 12-pulse or HF generators) (see chapter 4). These differences are important for selecting tube voltages in radiological imaging, and are listed in the exposure tables (see section 9.1.6).

9.1.1 Generator data

The X-ray generator supplies X-ray tubes with

- tube voltage U_R (kV) to generate the X-rays
- filament current I_R (mA) to generate the tube current
- stator voltage U_{St} and the drive frequency for driving the rotating anode (rotations per second in Hz)

The electrical output of the generator is calculated from the product of the tube voltage U_R (kV) and filament current I_R (mA) and is indicated in kilowatts (kW). To precisely calculate the output, the ripple of the tube voltage (see chapter 4) is multiplied by a dimensionless factor. With 1 and 2-pulse generators, this factor is 0.74. For 6-pulse generators, it is 0.95. It is around 1 for 12-pulse or multipulse generators.

Generator operating range

In diagnostic X-ray imaging, only radiation relevant to the image should be used given safety considerations. The X-ray generator must quickly provide the desired radiographic voltage (tube voltage) U_R at the required time, and immediately turn it off when the dose has been reached to provide a properly blackened image. Automatic controls and control loops have been accordingly developed for various applications such as automatic exposure control for spotfilming or automatic dose rate control (ADR) for fluoroscopy-guided examinations and interventions.

The operating range of a generator (Fig. 9.2) illustrates the relationship or dependence of the precise radiographic voltage U_R (kV), tube current I_R (mA) and the output P (kW) producible by the generator.

9 X-ray Imaging Technology

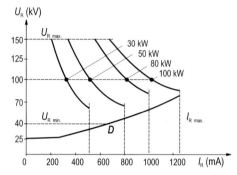

Figure 9.2 Definition of generator output according to IEC and DIN: The generator output P (kW) = k·radiographic voltage U_R (kV)·tube current I_R (mA), where k is a constant for the residual ripple of the radiographic voltage; approximately 1 for 12-pulse and multipulse generators.

The output ranges are divided into a low-kV range: 25 kV to 49 kV; normal range: 40 kV to 120 kV; high-kV range: 110 kV to 150 kV (see also chapter 4). See chapter 7 for information on the space-charge effect.

Generator output

The electrical output P (in Watts, W) is defined as the product of U (voltage in Volts, V) · I (current in Amperes, A).

- An output of 1 Watt exists when a current of 1 Ampere flows at a direct voltage of 1 Volt.

- Amperes as the unit of measure of electrical current are defined as a charge (the emitted electrons in the X-ray tube) that flows per second (from the cathode to the anode in this instance).

In radiological diagnostics, generators are used with outputs of approximately 1 kW to 150 kW (1,000 to 150,000 Watt). The performance data are standardized.

The higher the radiographic voltage U_R, the greater the ability of the resulting X-rays to penetrate (harder). In addition, the amount of scatter radiation that strikes the image receptor is also higher (see chapter 4 and section 9.3).

The maximum tube current I_R is approximately 1,200 mA, and the minimum tube current I_R is approximately 0.1 mA (with fluoroscopy). The tube current I_R is determined by the maximum output P of the generator or the type of X-ray tube.

Definitions

The performance data for X-ray generators is usually indicated by citing the manufacturer name followed by numbers and combinations of letters such as Polydoros™ 50 SX (Siemens AG). The letter combination indicates the area of application of the generator as defined by the manufacturer, and the number (50 in this instance) stands for a maximum generator output of 50 kW.

Fig. 9.3 classifies the generators according to their output. A generator that, for example, can operate an X-ray tube at a tube current of I_R at a maximum of 500 mA and tube voltage of U_R of 100 kV delivers an output P_R of 50 kW and is termed a 50 kW generator ($P = U \cdot I = 500$ mA $\cdot 100$ kV $= 50$ kW).

The standardized classification of generator output according to IEC 601-2-7 is as follows (e.g., for a 50 kW multipulse generator):

- 800 mA at 60 kV
- 500 mA at 100 kV
- 400 mA at 125 kV
- 320 mA at 150 kV

for a 80 kW multipulse generator:

- 1000 mA at 80 kV
- 800 mA at 102 kV
- 640 mA at 125 kV
- 530 mA at 150 kV

A power connection with a specific minimum internal impedance is required to provide high generator outputs for X-ray tubes. Among other things, the electrical lines must have a large diameter at the point at which they connect

Output	Main application
< 1.5 kW	Dental
1.5 to 15 kW	Mobile C-Arm: operating room
2.5 to 30 kW	Mobile generators (wall outlet operation): skeleton, bed exposures
25 to 35 kW	Mammography exposures
30 to 65 kW	Total routine X-ray diagnostics: workstations, fluoroscopy systems
50 to 80 kW	Total routine X-ray diagnostics: workstations, urology fluoroscopy systems, angiography
80 to 100 kW	Universal angiography, angiocardiography, neuroradiology

Figure 9.3 X-ray generators are classified in kW performance ranges corresponding to their diagnostic use

to the X-ray generator. Low-output mobile generators can run off of any mains socket outlet.

Exposure is understood as the product of the exposure strength (lux in Latin; the unit of exposure strength), and the duration of exposure in seconds. The term "exposure" was taken from photography since the processes of exposing film with light and film development are comparable (exposure machine, automatic exposure timer, exposure meter, exposure data, etc.).

In radiology, the exposure data that the X-ray generator provides are the tube current I_R in Amperes (A, usually mA), the tube voltage U_R in Volts (V, usually kV) and time t, or exposure time in seconds. These data are required to produce an X-ray exposure that can be used for diagnosis. They are termed exposure data or imaging parameters.

There are various techniques for setting these three imaging parameters. They are categorized as the 3, 2, 1 or 0-point techniques.

3-point technique

The term 3-point technique is used when all three parameters, mA, kV and s, are set manually. It is the exception and is used for imaging without an automatic exposure control, i.e., when the exposure time (s) has to be set separately.

2-point technique

In most generators, the tube current and exposure time are summarized as the mAs product. Only the two parameters kV and mAs are set in the 2-point technique.

Depending on the generator output, 50 mAs can be used by a generator control loop as follows:

- 50 mA · 1 s = 50 mAs (typical for generators with a small output)
- 200 mA · 0.25 s = 50 mAs (typical for generators with an average output)
- 500 mA · 0.1 s = 50 mAs (typical for generators with a high output)

When setting the mAs product, the set mAs is divided into an automatically set tube current (mA) and exposure time (s) by the generator processor so that the tubes are not overloaded and the operating times remain as short as possible.

1-point technique

If the exposures are created using an automatic exposure control, the automatic exposure control determines the right shut-off time. Only the kV is adjusted in the 1-point technique.

0-point technique

The generator automatically sets the necessary radiographic parameters from the exposure values in fluoroscopic imaging (see section 9.1.3). No parameters have to be set in the 0-point technique.

Constant and initial load

The performance of a generator determines if exposures are made with a constant load or falling load. If the tube current I_R of the X-ray tubes:

- can be maintained during the exposure time, the term *continuous load* is used
- is reduced in steps over the time of exposure, the term *stepped falling load* is used
- is continuously regulated based on the load characteristic of the respective X-ray tube, the term *continuously falling load* or *initial load* is used.

The "radiographic heating value" is reduced by constantly maintaining the focus temperature at its maximum permissible value (kW) over the entire exposure time. When automatic imaging is used, this optimizes the exposure time and minimizes motion unsharpness.

Fig. 9.4 shows the typical characteristic of the tube current I_{R1} (mA) with a falling load that is very high at the start of imaging (A). This value is maintained until the load curve of the X-ray tube is reached (B). Then the mA is reduced corresponding to the permissible load curve (b) until exposure is terminated (B). Exposure is shut off either by the set mA or the automatic exposure control.

When the load is constant, the tube current I_{R2} (A1) is lower and the exposure time is correspondingly longer (C1).

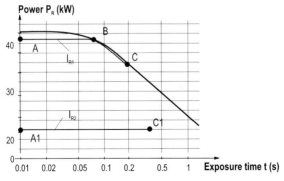

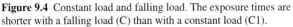

Figure 9.4 Constant load and falling load. The exposure times are shorter with a falling load (C) than with a constant load (C1).

Transparency compensation

Transparency is the ratio of the X-ray quanta that enter an object to those that exit the object (absorption). The transparency depends both on the thickness and density of an object.

Figure 9.5 Operating panel for generator parameters

An automatic control assigns the exposure data programmed for a "normal patient." When the patient thickness deviates substantially, an adjustment to the programmed data can be made by pressing transparency compensation buttons (Fig. 9.5, arrow). The transparency compensation affects the dose, the mAs and kV, and hence influences brightness and contrast.

Continuous fluoroscopy

In fluoroscopy, a specific dose is set with filament current I_H for tube currents I_R from 0.1 to 4 mA (in special cases up to 20 mA). In continuous fluoroscopy, the X-ray radiation is provided during the entire radiological examination at specific voltages and currents. Voltages and currents are summarized by the term "operating range," such as from 40 kV/0.2 mA to 110 kV/4.1 mA.

The automatic dose rate control calculates the data for an image from the ongoing fluoroscopy (see 0-point technique and automatic dose rate control).

9.1.2 The mAs product

The major factor in the exposure of an X-ray image is the radiation quantity (flux density, dose) over time (duration of exposure in seconds). The undefined concept of "radiation quantity" is sometimes used in the literature for the dose or the mAs product.

The mAs product (mAs = milli-Ampere seconds) is the product of the tube current I_R (mA) and the duration of exposure t (s), or the time of irradiation in seconds.

In targeted exposures (see chapter 7) in film-screen systems for example, the quantity of X-rays depends on the exposure, i.e., the intensity of the tube current I_R (mA as a measure of the tube current strength) and the time t (s) of an exposure:

9.1 Radiographic Parameters

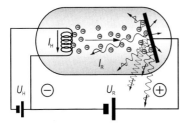

Figure 9.6 Schematic diagram of the filament circuit and the tube current I_R: The tube current I_R results from the filament current I_H of the directly-heated cathode in the potential gradient of the tube voltage U_R.

$$\text{mAs} = I_R \text{ (mA)} \cdot t \text{ (s)}.$$

The tube current I_R is a function of the tube voltage U_R with the filament current I_H as a parameter (Fig. 9.6). Each pair of values I_R and U_R is assigned a filament current I_H.

In fluoroscopy (dynamic radiographic imaging), a specific dose is set with filament current I_H for tube currents I_R from 0.1 to 4 mA (in special cases up to 20 mA).

When X-rays images are being made, the tube currents (up to 1,200 mA) during the exposure time $t(s)$ are kept constant (constant load), or are reduced by regulating a maximum value according to a predetermined characteristic (continuously falling load, initial load).

If the mAs product is increased, the overall blackening of the image increases, and the patient's radiation exposure rises.

- Changing the mAs product only influences the flux density of the X-rays and not the radiation quality. With thick objects, a greater flux density is required than with thin objects, i.e., the mA or mAs product must be larger to achieve a diagnostically equivalent result.

For an X-ray image, the tube voltage U_R is preset for the desired image character (contrast), and the size of the focus is preset for the detail recognition. The focus determines the permissible output P (kW) from which the tube current I_R (mA) can be determined as a function of the tube voltage U_R (kV) (for more on the focus, see section 7.3).

9.1.3 Automatic exposure controls

Automatic exposure controls measure the intensity I_ψ (energy flux density of the X-ray quanta, the dosage) of the X-rays between the object and the image receptor system, and send the measured values for control purposes to the CPU (see also Fig. 7.2).

For radiography, the automatic exposure control is used as an automatic dose-rate control (ADR) for fluoroscopy. The automatic exposure control terminates imaging as soon as the right X-ray dose for optimum blackening is attained.

Automatic exposure control is used in fluoroscopy with image intensifiers or flat detectors, and it maintains a constant dose rate at the image intensifier or detector input to provide optimum images on the monitor.

The recorded intensity measurement signal is used by:

- the automatic exposure control as a shut-off signal to terminate exposure,
- the automatic dose-rate control to regulate tube voltage U_R and tube current I_R.

Automatic exposure control with an ionization chamber

Automatic exposure controls (Fig. 9.7) are normally used in conjunction with an ionization chamber (see chapter 4, Measuring X-rays), or with a semiconductor detector such as in mammography.

The automatic exposure control uses the accumulated dose to calculate the measured values for regulating the shut-off time for optimum image blackening. Only the kV (tube voltage U_R) must still be set. The automatic exposure control turns off the exposure when the mAs required for optimum blackening of the X-ray image is reached (1-point technique).

The X-rays coming from the patient generate a very small ionization current (approximately in the picoampere range, $1\ pA = 10^{-12}\ A$) in the dose measuring device (ionization chamber) to which a voltage (potential) of a few 100 V

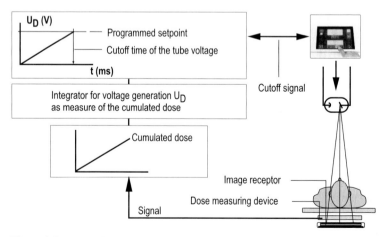

Figure 9.7 Automatic exposure controls

is applied. After appropriate amplification (the first amplification stage is integrated in the ionization chamber housing), the processors of the automatic exposure control (in the generator) compare the current with a saved nominal value. As soon as this value is reached, i.e., the dose for the optimum density, the automatic exposure control sends the signal to the X-ray generator to stop imaging.

Each X-ray image contains an area that is relevant for diagnosis. This area is known as the "dominant" area. The following holds true for the automatic exposure control: When the blackening in the dominant area is correct, then the image is illuminated well overall. The automatic exposure control therefore concentrates only on properly exposing the dominant areas selected in the generator operating panel. Many ionization chambers contain several measuring chambers. For each image, the measuring chamber (dominant) is used that lies in the area of the object to be recorded (see Fig. 9.12, selective dominant measurement).

For example, the 3-field chamber with three measuring fields (dominants) is frequently used. Before each exposure, the right measuring field or dominant must be set.

When exposures are taken with automatic exposure control (1-point technique), first the kV (tube voltage U_R) must be entered (such as 81 kV). When film-screen systems are used for imaging, the required dose largely depends on the sensitivity of the intensifying screens (see chapter 8). Hence the sensitivity class H, U or D must be entered. The blackening compensation (Fig. 9.8, top middle) can be used to generally adjust the shut-off dose upward (+1, +2, etc.) or downward (−1, −2, etc.). The setting refers to the exposure points. It means that the shut-off dose has been changed plus (+) or minus (−)1, 2, etc. points in comparison to the standard setting (for information on exposure points, see section 9.1.6). When ionization chambers have several measuring fields, the corresponding dominants must be entered (Fig. 9.8, bottom right).

Ionization chambers are nearly shadow-free and are arranged in front of the image receptor system. Semiconductor radiation detectors (see chapter 8, Flat Detector Systems) have much higher absorption than ionization cham-

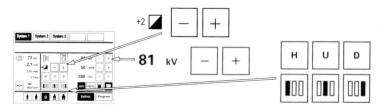

Figure 9.8 Settings when automatic exposure controls are used

bers. Since they are not shadow-free, semiconductor radiation detectors must hence be placed behind the image receptor system.

An automatic exposure control is usually used in the form of a dose measurement chamber for X-ray systems that combine radiography and fluoroscopy. The sensor measuring field can be the IONTOMAT ionization chamber or a semiconductor radiation detector as used in mammography.

Organ programs, programmed radiographic technique

All of the relevant exposure data and imaging parameters are archived in the organ programs for specific organs. All of the parameters can be modified according to individual requirements. The organ-specific settings are retrieved by selecting the corresponding organ programs; this method is also termed the "programmed radiographic technique" (Fig. 9.9).

The automatic exposure control is optional in the programmed radiographic technique.

If the automatic exposure control is not used (which it rarely is today), the organ program for the "normal patient" saves the kV and mAs. Transparency compensation must be manually selected for patients who strongly deviate from the norm such as adipose (obese) patients. The transparency compensation keys on the operating panel are used for this purpose (see Fig. 9.5). Generally, one step corresponds to one exposure point (see section 9.1.6 for information on exposure points).

Organ programs that are programmed, unprogrammed, newly programmed or must be reprogrammed are retrieved by the program key (Fig. 9.9-P). Only the kV for normal patients is saved in the organ programs when automatic exposure controls are used. The automatic exposure control switches the mAs. Transparency compensation should also be used for patients strongly deviating from the norm (Fig. 9.5). The kV is changed for the object.

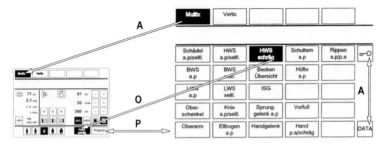

Figure 9.9 Programmed radiographic technique

Usually the keys for the object program are programmed for a specific organ, in this case "angled cervical spinal column" (Fig. 9.0-O). Hence a key can be programmed for the same imaged object several times, i.e., for the normal technique, indirect technique and slices (depending on workstation, Fig. 9.9-A). The automated system automatically assigns the correct exposure data.

Most generators have an override function. This enables you to correct the exposure data even after selecting the relevant program.

Automatic exposure control through the automatic dose-rate control

The automatic dose rate control (ADR) is used in X-ray systems that offer fluoroscopy. In contrast to automated exposure control with an ionization chamber, the dose rate in front of the image receptor is maintained at a constant value in fluoroscopic systems (Fig. 9.10).

The intensity of the radiation is measured after the object as is the case with automated exposure controls for X-ray images. This measured value is used for automatic dose rate control (ADR) of the exposure. The measured value representing the dose rate in the image receptor is evaluated by the automatic systems. They generate control signals for the generator.

A photomultiplier functioning as a measuring field sensor (dose measuring device) continuously detects the dose rate indirectly via the optical brightness that is proportional to the dose rate (Fig. 9.10). A control loop keeps the dose rate constant independent from the object. The ADR supplies signals to control the tube voltage U_R (kV) and the tube current I_R (mA).

If the present measured value (actual value) is too low in comparison to the nominal value if, for example, the object is too thick, the ADR increases the tube voltage (kV) and tube current (mA) until the target value is reached. If a difference from the value arises, for example, from rotating or moving the patient, the control is reactivated. The control speed and precision of the ARD are so high that the brightness of the monitor image is not influenced by ongoing adjustments.

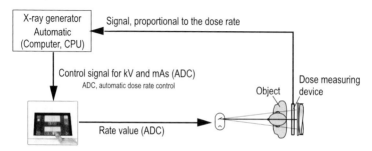

Figure 9.10 Automatic dose rate control (ADR)

Dose rate characteristic

Either the tube voltage U_R (kV) or the tube current I_R (mA) or preferably both parameters together must be continuously changed to continuously adapt the radiation intensity to the just-irradiated object. The manner in which the kV and mA are changed in this control system depends on the fluoroscopy characteristic or the dose-rate characteristic.

Depending on whether the examination is application-related or object-related, any number of kV-mA links can be individually set and programmed, i.e., each kV is assigned an mA (Fig. 9.11).

- Antiisowatt characteristic
  The standard characteristic (S) shows the most frequently used control characteristic. This is also called the "antiisowatt characteristic." It represents a compromise between the highest contrast and lowest radiation exposure.
- Dose-reduced
  Characteristic "R" is a dose-reduced characteristic. By means of a quickly rising kV, it ensures a minimum dose and is preferably used in pediatrics, gastrointestinal examinations and routine operation.
- High contrast
  Characteristic "H" is for a low kV and high contrast such as in vascular examinations using contrast agent. A slowly rising kV and higher mA values are attained with this characteristic. The goal is to obtain the highest contrast, for example, in intervention-related examinations. The additional reduction in the rising kV (a) is due to the need for maximum iodine contrast. Upon reaching the limit load (b) during continuous fluoroscopy, the characteristic reverses.

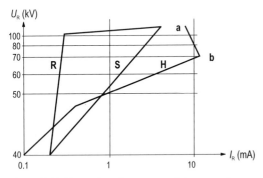

Figure 9.11 Examples of programmed dose rate characteristics for the ADR

9.1 Radiographic Parameters

Selective dominant measurement (SDM)

When high quality is required especially in digital radiography, the initial brightness of the image intensifier is detected in a semiconductor sensor set up in a diode matrix (Fig. 9.12). The sensor has a left (L), middle (M), right (R) and an alternative central (C) measuring field that are also termed dominants.

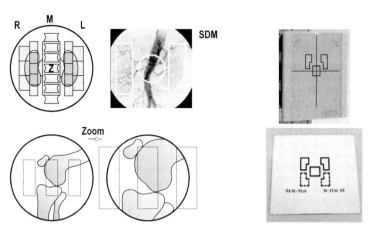

Figure 9.12 Illustration of selective dominant measurement (left) and an arrangement of the measuring fields at the bucky tray of a bucky wall stand for an automated exposure control. Underneath are 3-field templates to be inserted in the accessory rails of the collimator since the dominants are covered when the patient stands in front. The outlines of the dominants on the patient are portrayed with the aid of the light localizer (see section 7.4).

The dominants can be selected individually or in any combination, similar to the measuring chambers (dominants) of an ionization chamber when automatic exposure controls are used for X-ray images. The measuring fields are archived in the organ program, i.e., no additional steps are necessary. The basic measuring field size is preset. The size can be changed in special instances; for example, it can be set to be smaller for pediatric examinations.

Examination-oriented selective dominant measurement (SDM) ensures the proper image characteristics during fluoroscopy and the desired exposure of the X-ray image. The measuring fields can be enlarged on the sensor using the zoom function (Fig. 9.12, bottom left). The object evaluation remains the same, and so do the image characteristics.

Transferring the kV from fluoroscopy values

The automatic dose-rate control (ADR) determines the transparency, i.e., the absorption behavior of the object, during fluoroscopy from the permanent control of the tube voltage U_R (kV) and the tube current I_R (mA). These continuously measured kV and mA values are used by the processor to calculate the best kV for the next exposure that may follow.

These exposure kVs are generally not identical to the fluoroscopy kVs. These values are calculated from the transparency according to "transfer characteristics" that are generally designed or set for specific organs. This gives rise to the expression "transfer the kV from fluoroscopy." When the exposure is triggered, the generator automatically and immediately sets the exposure kV that is precalculated during fluoroscopy. Since the automatic exposure control also turns off the exposure, i.e., ensures the optimum mAs, no exposure data need to be set before taking the X-ray in this case. This particular setting technique is also termed the 0-point technique (see section 9.1.1).

Controlling automatic exposure in serialography

In serialography, there is not enough time to cool the anode between the individual exposures. To keep the tubes from overloading, the overall number of exposures must be limited. In modern generators, the exposure frequency, scene duration and dose per image must be entered. The automatic control saves all the data that refer to the load capacity of the tube and radiation head such as the heating and cooling curves (see section 9.1.5). During fluoroscopy-guided catheterization, the processor calculates the series exposure data from the transparency, i.e., the absorption behavior of the object.

Automatic exposure control during conventional tomography

In addition to the kV setting, the correct mAs product is also relevant in conventional tomography (see section 9.4). Since the operating time is predetermined by the tomographic angle, the automatic exposure control does not regulate the exposure time, but rather sets the tube current (mA). It calculates the mAs necessary for optimum exposure from the ionization current in the first milliseconds of imaging. This mAs is divided by the set operating time of the planigraphic procedure to calculate the tube current I_R (mA) which is then automatically set.

When exposures are taken with a constant source-image distance (SID), this is the value for optimum exposure. Since in conventional tomography the slices are taken in horizontal planes, i.e., the focus and imaging system are parallel and opposing, the SID does not remain constant. The automatic exposure control compensates by correcting the tube current (mA).

The different manner of calculating the data for the automatic exposure control is expressed by the information in data sheets, for example, with the automatic exposure control IONTOMAT™ or IONTOMAT-P™ (P stands for planigraphy).

9.1.4 Error display and data logs

Processor-controlled X-ray generators periodically check the functioning of all generator circuits such as the high-voltage circuit or filament circuit, the rotating anode drive system, the automatic exposure controls, etc., using an automatic, self-diagnosing test program. The test program identifies any arising errors and displays them as a number code. All relevant operating data are saved for technical support.

All of the data are processed by the generator using processor-controlled computing programs. Data logs of the exposure data, data on the calculated dose-area product, the fluoroscopic time or the tube load computer, can hence be saved, displayed, printed out or sent to other computers via networks.

9.1.5 Tube data

In addition to the described generator data, other physical parameters play a major role in radiography in creating a high-quality image that allows the radiologist to make a reliable diagnosis. The nature of the focuses and the thermal properties of the X-ray tube are important parameters. The design and function of the X-ray tube are described in chapter 7.

Focus

Stationary anode tubes are usually manufactured as "single focus tubes." Rotating anode tubes are chiefly built with two spiral-wound filaments as "dual-focus tubes." The user can therefore choose between two focus sizes (Fig. 9.13). The smaller focus (typically 0.6) generates less focal unsharpness. It is always used when its lower load capacity (output) is sufficient. In all other cases, a large focus (typically 1.0) is used. Three-focus tubes usu-

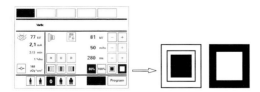

Figure 9.13
The focus size can be selected depending on the required imaging technique

ally have a third focus, a "microfocus." It is usually 0.3 or less, and is used for enlargements such as in cardioradiology or neuroradiology since it has less focal unsharpness.

The nominal focus values, permissible deviations of nominal values and the measuring methods are standardized according to IEC and DIN. The width and length are dimensionless values only indicated by a number (see Fig. 7.15, section 7.3).

The focus unsharpness is a function of the focus size or optical focal spot (see Fig. 7.14, section 7.3). It influences attainable image sharpness (see section 9.3). The influence of the anode angle is also discussed in section 9.3.

Focus jump

When both filaments in a two-focus tube are arranged sequentially along the focus path, they are termed "superposed focuses." If the two spiral-wound filaments are adjacent to each other along the focus path, a focus jump arises when switching from one to the other focus. The focus jump produces a slight change in the exit angle of the central ray that is irrelevant to the scanning geometry.

Focal output

The larger the dimensions of the focus, the more load it can assume. In addition, its geometric unsharpness or focal unsharpness also increases (see section 9.3).

As with X-ray generators, the output of X-ray tubes is indicated in kilowatts (kW), i.e., the product of the tube current I_R (mA) and radiographic voltage U_R (kV) (see section 7.2). The focal capacity of a rotating anode tube is, for example, identified as a 40 kW focus when it can handle an output of 40 kW for 0.1 seconds. 1 second is used when defining the focal output of stationary anodes.

Definitions

The performance data of X-ray tubes are usually indicated by citing the manufacturer name followed by numbers and combinations of letters. The numbers indicate the maximum tube voltage U_R and the maximum load of an X-ray tube or the permissible focal load.

Example: OPTITOP™ 150/40/80 HC-100.

150	indicates the maximum tube voltage U_R, i.e., the maximum radiographic voltage in kV that can be set for the generator
40	the maximum load of the smaller focus (such as 0.6) is 40 kW, i.e., the load limit before the anode is overheated or damaged

80	the maximum load of the larger focus (such as 1.0) is 80 for the above-cited reasons; kW
HC	are usually the manufacturer's internal designations of groups or features
100	indicates the diameter of the rotary anode; in this instance, 100 mm (others are 120 or 125, usually in high-performance tube)

Some manufacturers indicate the short-term loads of focuses as a function of the preliminary load. For example, 54 kW can be attained for a 40 kW focus when the preliminary load is 0 Watts, or 40 kW can be attained with a preliminary load of 300 Watts. A tube should always be started up with a certain preliminary load for thermal and mechanical reasons.

Tube nomogram

X-rays are taken as quickly as possible, i.e., the anode of the X-ray tube is only activated for a very brief time (short-time load). The heat generated along the focus path, i.e., on the surface of the anode, has almost no time to flow into the anode disk during such individual exposures.

The thermal threshold can be reached more quickly during serial radiography. The threshold is reached when the temperature of the focus path rises to a maximum of approximately 2,000°C. The larger the focus and the higher the number of anode rotations, the higher the load threshold.

Tube nomograms (focus load curves) indicate how long the focus can or may undergo a load at a specific tube voltage U_R (kV) at what maximum tube current I_R (mA).

The representation of the tube nomogram on the left in Fig. 9.14 is preferred when there are questions that primarily deal with the tube voltage U_R and tube current I_R. The load curves of a nomogram are derived from respective parameters such as:

- the type of high-voltage waveforms (the tube voltage of a 12 pulse or HF generator in this instance)
- maximum possible focus load (40 kW in this instance)
- assigned focus size (0.6 in this instance)
- anode speed (8,500 rpm in this case)

The thermal state of the tube can be monitored by directly measuring the anode temperature.

Example: Given a tube voltage U_R of 90 kV and tube current I_R of 280 mA, the tube load threshold is reached after a maximum of 1 seconds (gray shaded area in Fig. 9.14, left).

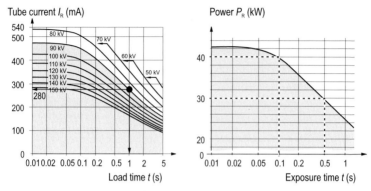

Figure 9.14 Illustration of a tube nomogram using the example of a 0.6 focus with a 40 kW load and an anode speed of 8,500 rpm (left). The exposure time as a function of the tube output P_R (right): for a 40 kW focus, the load threshold is reached after 0.1 seconds. If only 30 kW is applied to the 40 kW focus, the load threshold is reached after 0.5 seconds.

Or: Every combination of tube current and exposure time, i.e., every mAs product (mAs values), is permissible that lies within the gray shaded area. Higher values would overload the tube. Overloading can be excluded since all generators are equipped with electronic overload protection devices.

Another example: Given a tube voltage U_R of 100 kV, a tube current I_R of 400 mA can flow 0.1 seconds without overloading. This is the range under the 100 kV curve (Fig. 9.14, left).

In the tube nomogram in Fig. 9.14, the representation on the right is preferred when there are questions that primarily deal with the tube or focus output P_R. This curve marks the load limit for a 40 kW focus.

Figure 9.15 compares the interdependence of the focus size and anode speed:

- A with B: Although the focus of B is larger than that of A, A can take a greater load due to the higher anode speed.
- A with C: Both focuses are nearly the same, but A can take a much higher load due to the higher anode speed.
- B with C: Given the same anode speed, B can take a higher load due to the larger focus.

D with C: Despite the larger focus, the load capacity of D is only 10% that of C since D is a stationary anode tube. Stationary anode tubes are only operated in connection with lower-output generators due to their lower load capacity. Given exposure times > 0.1 sec., mAs products are obtained that are too low for practical purposes.

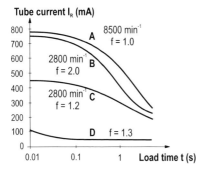

Figure 9.15
Comparison of the interdependence of the focus size and anode speed. The load capacity in the short-time range rises with the focus size (f) and the anode speed (rpm).

Heating and cooling curves

In angiograms, interventions and dynamic computer tomography, the rotating anodes of the tubes are subject to an extremely high load or a long-time load.

The heat generated along the focus path spreads to the anode disk. The temperature in the anode disk rises until the disk starts to glow dark red. The additional load on the anode disk depends on the heat storage capacity of the X-ray generator. Criteria such as the focus size and anode speed that were relevant for the short-time load are no longer relevant. The heat storage capacity and hence the properties of the long-time tube load chiefly depend on the material and design (including the weight) of the anode disk (see section 7.3).

The anode must be able to radiate large amounts of heat. The radiated heat increases to the 4th power of the temperature of the anode disk. The term high-temperature radiation cooling is also used. Black bodies radiate most effectively so that graphite is optimum for high-performance tubes.

The inside of the tube housing that includes oil, the stator, etc., is heated by the heat radiated from the glowing rotating anode. The heat is discharged more slowly, however, than the supply of heat during intensive use. For this reason, the thermal behavior of the overall X-ray generator must be monitored in addition to protecting the anode disk from overheating. This is illustrated in heating and cooling curves (Fig. 9.16).

The heat storage capacity of an X-ray generator indicates the energy that the anode can absorb. As mentioned previously, Joule (J) equals the unit of energy $E = 1$ eV $= 1.602 \cdot 10^{-19}$ J (1.602 ... is the elementary charge of an electron). The heat storage capacity is indicated in Joules (J) or heat units (HU).

- 1 J = 1.35 HU
 This is not to be confused with Hounsfield units (HU) (see appendix).

9 X-ray Imaging Technology

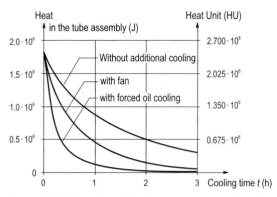

Figure 9.16 Illustration of the time characteristic of cooling a heated tube. The cooling time can be reduced substantially by intentionally drawing off the heat.

At the moment, X-ray tubes for fluoroscopy and angiography systems are being made with up to 2.5 MHU, and tubes are being created with up to 8 MHU for computer tomography.

The function and design of the Straton™ X-ray tube is described in section 7.3. The described nomograms do not apply to these tubes. Given the special construction, i.e., a rotating anode directly connected to the rotating tube housing, the heat can be removed much more effectively. Direct cooling enables cooling rates of 4.7 MHU/min. and eliminates the need for a large heat storage capacity.

Tube load computers

Tube load computers protect the connected X-ray tubes from overloading during imaging. It calculates the required pause time from the set exposure data and the physical and geometric properties of the X-ray tube.

The X-ray tubes of modern generators do not overload from single exposures or several exposures taken at long intervals. The tube loadability threshold can be reached from a quick sequence of exposures such as in serialography. Tube load computers calculate the temperature rise for each load and also the ongoing cooling of the anode disk. The calculation is based on the saved heat and cooling characteristics of the utilized X-ray tube. The current temperature of the anode disk is displayed on the generator operating panel (Fig. 9.17). The tube load calculator uses the calculated current temperature of the anode disk and the set data for the next exposure to calculate the pause that is required to trigger the next exposure.

Switching the tube load (right):

The display "80% (kW) 100%" (Fig. 9.17) refers to the possibility that exists with many generators of switching between two tube loads. When the tube

9.1 Radiographic Parameters

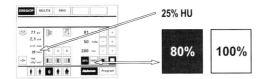

Figure 9.17 25% HU refers to the current temperature of the anode disk

load is set to 80%, the exposures are taken with less tube current (mA) and longer exposure times (s). The focal path therefore does not reach such high temperatures as a 100% setting. The tube is subject to a lower load.

A 100% setting is, however, permitted. It generates the values of the stipulated exposure data. When the tube load computer demands (displays) a pause at 100% tube load, it can generally be shortened by switching to 80% mode since switching reduces the tube load. In medical interventions, the tube load computer does not block imaging. It is up to the user to evaluate the pause request.

9.1.6 Exposure tables

X-ray systems without automatic exposure control must be used in "free operation." The exposure parameters such as kV, mA and s or mAs are set by referring to exposure tables or based on the extensive experience of the investigator.

Exposure tables contain basic values for standard conditions such as the object density, SID, film-screen system, grid type, etc. If these standard conditions are not met, the indicated data must be correspondingly recalculated.

Siemens point table

The Siemens point table has proven to be valuable in recalculating exposure data (Fig. 9.18, 9.19 and 9.20). It allows an adaptation to different exposure conditions by adding or subtracting "exposure points" (EP or simply points). The following rules are used:

- 1 EP changes the exposure 25%; i.e., 1 EP compensates for an object density of 1 cm (with the exception of the lung) in reference to the exposure.
  Or:
  1 EP (+) increases the dose by a factor of 1.25, and 1 EP (−) lowers the dose by a factor of 0.8.

- 3 EP double or halve the exposure; i.e., 3 EP compensate for an object thickness of 3 cm (with the exception of the lungs) in reference to the exposure.
  Or:
  3 EP (+) increase the dose by a factor of $1.25 \cdot 1.25 \cdot 1.25 = 1.95 =$ approx.

2.0 and
3 EP (–) decrease the dose by a factor of $0.8 \cdot 0.8 \cdot 0.8 = 0.512 =$ approx. 0.5.

No theory is required for recalculation. It is sufficient to become briefly familiarized with the three applicable tables. These are:

- the basic exposure table (Fig. 9.18),
- the conversion table for kV and mAs (Fig. 9.19)
- the correction table (Fig. 9.20) for screens, grids, image receptor focus, distance, etc.

The basic exposure table provides the required exposure in EPs for each cited object under the cited conditions (SID, screens, generator, etc.).

Example: For an a.p. X-ray of a skull, 27 EPs are indicated in the basic exposure table (Fig. 9.18). In the kV and mAs column, 77 kV and 16 mAs are indicated for these 27 EPs. In the conversion table (Fig. 9.19), 77 kV corresponds to 15 EPs, and 16 mAs corresponds to 12 EPs.

These 27 EPs could also be obtained according to the conversion table (Fig. 9.19) by different combinations of kV/mAs such as 73 kV (14 EPs) and 20 mAs (13 EPs), or 81 kV (16 EPs) and 12.5 mAs (11 EPs). Since, however, kV changes influence the contrast, the kV is generally not changed from the value in the basic exposure table. An exception is made when there is a particularly large deviation from the standard object thickness. In this case as well, it is recommendable to keep the kV change as small as possible.

For mAs, the conversion tables clearly show (Fig. 9.19) that each 3 EP change doubles or halves the dose. *Example:* 10 BP = 10 mAs, 13 BP = 20 mAs.

A different rule applies for kVs in the conversion table (Fig. 9.19). At a very low kV, an increase of a few kV (such as from 40 to 44 kV) doubles the dose (mAs), i.e., causes an increase of 3 EPs. In the middle kV range, approximately 10 kV is required to double the dose, for example, 60 kV correspond to 10 EP, 70 kV correspond to 13 EP, and 81 kV correspond to 16 EP. As the kV range increases, the kV change must be accordingly higher to change the dose. The reason is that the attenuation coefficients (of tissue for example) in the higher kV range change less than in the very low kV range (see section 3.3).

Basic exposure tables

The exposure guidelines in the basic exposure table (Fig. 9.18) apply to optimum developing conditions and the suggested film-screen sensitivity classes when using multipulse or 12-pulse generators. In the case of deviating exposure parameters, the necessary changes to the exposure data must be determined in the correction and conversion table (Fig. 9.20 and 9.19).

9.1 Radiographic Parameters

Skull		Thickness (cm)	SID (cm)	Sensitivity class	Grid (Pb 12/40)	Points	kV	mAs
Skull survey	p.-a./a.-p.	19	115	400	with	27	77	16
Skull survey	lat.	16	115	400	with	26	73	16
Skull survey	axial	22	115	400	with	32	85	32
Petrous bone	sag.	17	115	200	with	30	73	40
Petrous bone, Stenvers		17	115	200	with	30	73	40
Nasal sinuses	p.-a.	22	115	400	with	29	77	25
Orifice of optical nerve, Rhese		17	115	200	with	29	77	25
Mandible	lateral	11	105**	200	without	17	57	6.3

Chest		Thickness (cm)	SID (cm)	Sensitivity class	Grid (Pb 12/40)	Points	kV	mAs
Ribs 1 - 7	p.-a./a.-p.	20	115	400	with	25	70	16
Ribs 8 - 12	p.-a./a.-p.	22	115	400	with	26	73	16
Sternum	p.-a.	21	115	400	with	24	70	12.5
Sternum	lat.	30	115	400	with	25	73	12.5
Clavicle	p.-a./a.-p.	14	115	400	with	21	66	8
Scapula	lat.	17	115	400	with	23	66	12.5
Lung	p.-a./a.-p.	21	180	400	with	26	125	2
Lung (in bed)	a.-p.	21	115	400	without	13	60	2
Lung, heart	lat.	30	180	400	with	29	125	4
Lung, heart (in bed)	lat.	30	115	400	without	16	60	4
Esophagus	obl.	28	70/115*	400	with	24/28	90	4/10

Abdomen		Thickness (cm)	SID (cm)	Sensitivity class	Grid (Pb 12/40)	Points	kV	mAs
Kidney, gallbladder	lat.	27	115	400	with	30	81	25
Kidney, gallbladder	a.-p.	19	115	400	with	27	73	20
Urinary bladder	a.-p.	19	115	400	with	25	77	10
Urinary bladder	axial	21	115	400	with	27	81	12.5
Content study of stomach	p.-a.	22	70/115*	400	with	27/31	109	4/10
Bulbus	p.-a.	22	70/115*	400	with	27/31	109	4/10
Gastrointestinal tract, survey		22	70/115*	400	with	27/31	109	4/10
Stomach, relief		22	70/115*	400	with	27/31	109	4/10

Pelvis		Thickness (cm)	SID (cm)	Sensitivity class	Grid (Pb 12/40)	Points	kV	mAs
Pelvis, hip	a.-p.	20	115	400	with	24	77	8
Sacrum, coccyx	a.-p.	19	115	400	with	32	90	25
Sacrum, coccyx	lat.	33	115	400	with	36	90	63

Figure 9.18 Basic exposure tables.
Note on the SID column: *) 70 cm for undertable radiographic systems, and 115 cm for overtable radiographic systems. **); for table exposures, i.e., when the cassette is positioned on the table directly on the patient. See chapter 9.4 for projection designations.

9 X-ray Imaging Technology

Spinal column		Thickness (cm)	SID (cm)	Sensitivity class	Grid (Pb 12/40)	Points	kV	mAs
Cervical vertebrae 1 - 3	oral	13	115	200	with	26	70	20
Cervical vertebrae 4 - 7	a.-p.	13	115	200	with	28	78	25
Cervical vertebrae 1 - 7	lat.	12	115	200	with	25	70	16
Cervical vertebrae 1 - 7	obl.	13	115	200	with	26	70	20
Thoracic vetebrae	a.-p.	21	115	400	with	31	77	40
Thoracic vetebrae	lat.	30	115	400	with	33	81	50
Lumbar vertebrae 1 - 4	a.-p.	19	115	400	with	32	77	40
Lumbar vertebrae 1 - 4	lat.	27	115	400	with	38	90	100
Lumbar vertebrae 1 - 4	obl.	22	115	400	with	33	85	40
Lumbar vertebra 5	a.-p.	22	115	400	with	32	90	25
Lumbar vertebra 5	lat.	33	115	400	with	36	90	63

Upper extremities		Thickness (cm)	SID (cm)	Sensitivity class	Grid (Pb 12/40)	Points	kV	mAs
Shoulder joint	a.-p.	11	115	200	with	23	66	12.5
Shoulder joint	axial	11	105**	200	without	20	66	6.3
Upper arm	a.-p./lat.	8	105**	200	without	18	60	6,3
Elbow	a.-p.	6	105**	200	without	15	57	4
Elbow	lat.	8	105**	200	without	15	57	4
Forearm	a.-p.	6	105**	200	without	14	55	4
Forearm	lat.	7	105**	200	without	15	55	5
Wrist	d.-v.	4	105**	200	without	12	46	6.3
Wrist	lat.	6	105**	200	without	15	52	6.3
Hand	d.-v.	3	105**	200	without	9	46	3.2
Hand	lat./obl.	6	105**	200	without	10	46	4
Finger		2	105**	200	without	7	46	2

Lower extremities		Thickness (cm)	SID (cm)	Sensitivity class	Grid (Pb 12/40)	Points	kV	mAs
Neck of femur	axial	22	155*	400	without	22	77	5
Femur	proximal	13	115	400	with	23	73	8
Femur	distal	12	115	400	with	21	66	10
Knee joint	a.-p.	12	115	200	with	22	63	12.5
Knee joint	lat.	10	115	200	with	21	63	10
Knee joint fissure		12	105**	200	without	15	60	3.2
Patella	axial	7	105**	200	without	15	60	3.2
Tibia	a.-p.	11	105**	200	without	14	60	2.5
Tibia	lat.	9	105**	200	without	13	60	2
Ankle	a.-p.	9	105**	200	without	14	57	3.2
Ankle	lat.	7	105**	200	without	13	57	2.5
Oscaleis	lat.	7	105**	200	without	12	55	2.5
Oscaleis	axial	10	105**	200	without	12	55	2.5
Matatarsus	d.-pl.	5	105**	200	without	11	52	2.5
Matatarsus	obl.	6	105**	200	without	11	52	2.5
Foot	lat.	7	105**	200	without	12	52	3.2
Toes		3	105**	200	without	9	48	2.5

Bild 9.18 Basic exposure tables *(continued)*

Point table rules

Point table rules

A change of 1 exposure point causes a change in optical density in the film of approximately 0.25. Since the eye scarcely recognizes differences in density less than 0.25, changes corresponding to less than 1 exposure point such as a transition from 70 kV to 68 kV or 40 mAs to 44 mAs have almost no effect. The kV and mAs gradations of most generators take this fact into account; i.e., the transitions generally correspond to one point, and sometimes one-half a point. This is somewhat comparable to the gradations of the exposure time and apertures of cameras (1/60 s, 1/125 s, 1/250 s, etc.) that are also dimensioned to cause a recognizable change in the film exposure.

Since 1 cm object thickness corresponds to 1 EP (with the exception of lungs), 3 EPs are required for an additional 3 cm, i.e., the dose must be doubled. The same holds true in the other direction. For an object that is 3 cm thinner, 3 EPs fewer are required, i.e., only the half dose.

Conversion table

The following table can be used to convert points into kV and mAs (Fig. 9.19).

Correction tables

The tables in Fig. 9.20 show the corrections in exposure points for deviations from the initial conditions. A difference of three exposure points causes the dose to double or halve in the film plane (image receptor plane). The difference of one exposure point causes a change in blackening Δ_s of approximately 0.25 on the X-ray film (image receptor).

Significance of the negative points: When changing to a screen of a higher sensitivity class or, for example, when a grid is not used, the required points fall as indicated.

kV	-	-	-	-	-	-	-	-	-	40	41	42	44	46	48	50	52
Points	-10	-9	-8	-7	-6	-5	-4	-3	-2	-1	0	1	2	3	4	5	6	7
mAs	0.1	0.13	0.16	0.2	0.25	0.32	0.4	0.5	0.63	0.8	1	1.3	1.6	2	2.5	3.2	4	5

kV	55	57	60	63	66	70	73	77	81	85	90	96	102	109	117	125	133	141
Points	8	9	10	11	12	13	14	15	16	17	18	19	20	21	22	23	24	25
mAs	6.3	8	10	12.5	16	20	25	32	40	50	63	80	100	125	160	200	250	320

kV	150	-	-	-	-
Points	26	27	28	29	30
mAs	400	500	630	800	1000

Figure 9.19 Table to convert points (EP) into kV and mAs

9 X-ray Imaging Technology

SID	cm	65	75	85	95	105	115	130	145	160	185	210	235	260	290	325	360	400
	Points	-5	-4	-3	-2	-1	0	+1	+2	+3	+4	+5	+6	+7	+8	-9	-10	-11

Screens			Generator		Grid	
Sensitivity	Points 50-90 (kV)	Points 90-150 (kV)	Type	Points	Type	Points
800	-7	-8	DC voltage	0	without	-6
400	-3	-4	Multipulse	0	Pb 8/40	-2
200	0	-1	12-pulse	0	Pb 10/40	-1
100	+3	+3	6-pulse	+3	Pb 12/40	0
50					Pb17/70	0

Object		Tomography		Zonography	
Condition	Points	Figure	Points	Figure	Points
thin	-3 ... -1	———	+2	———	-1
thick	+1 ... +3	◯	+3	◯	-1
Close collimation	-2	⌒	+3		
Plaster half-shell	... +3	◎ 16°	+3		
Dry plaster	... +5	◎	+4		
Wet plaster	... +7				

Figure 9.20 Correction table for converting to a different SID, a different screen sensitivity, a different generator, etc.

General instructions

The following method for using the point table has proven to be helpful:

- Determine which conditions of the present exposure are different from those of the exposure table.
- How many points (from the table in Fig. 9.19 and 9.20) more or less correspond to this deviation?
  Ask yourself: is this an exception?
  (for lungs, deduct approx. 3 cm = 1 point; for children between 6 and 16: deduct another point)
- Decide: what to change, kV, mAs or both?
  Find the new value from the tables (9.19 and 9.20). Make sure that the calculated points are used in the right direction, i.e., added or subtracted.

To calculate the new exposure when there are changes in the SID, use the correction table (Fig. 9.20), and the following table (Fig. 9.21).

SID (cm)	Change: SID (cm)				
	70	100	115	150	200
70	1	2	2.7	4.6	8.2
100	0.5	1	1.3	2,35	4
115	0,.37	1.3	1	1.7	3.32
150	0.2	0.45	0.58	1	1.75
200	0.12	0.25	0.33	0.55	1

Figure 9.21 Factors to consider in order to change the mAs product when there are changes to the SID. Other distances can be calculated using the inverse square law (section 3.6).

The following three examples illustrate typical uses of the point table.

Recalculating mAs for other object densities

Take the example of a p.a. chest X-ray, SID 180 cm, intensifying screen: 400, scatter radiation grid: Pb 12/40. The object thickness of 24 cm is +3 cm above the value in the table. For lungs, a difference in thickness of 3 cm approximately corresponds to 1 point. An object thickness greater by 3 cm hence means that the exposure must be increased by one point. The kV should not be changed since this would affect the contrast. For this reason, the mAs is increased.

The data in the table (Fig. 9.18) for this exposure are 125 kV and 2 mAs (at 26 points). Corresponding to the table (Fig. 9.19), 23 points are assigned for 125 kV, and 3 are assigned for 2 mAs. Increasing the mAs by 1 point yields 4 points, i.e., 2.5 mAs. The settings for the exposure are hence 125 kV and 2.5 mAs.

Recalculating the mAs when changing the tube voltage

Take the example of an a.p. kidney X-ray: The table (Fig. 9.18) indicates 73 kV and 20 mAs (at 27 points). The exposure should be taken at 77 kV, however, i.e., the increased kV must be compensated by lowering the mAs. How much does the mAs have to be reduced?

Table 9.19 shows a kV increased by 1 point when changing from 73 kV (14 points) to 77 kV (15 points). Since the total points should not change (27 points), the mAs should be reduced 1 point. 20 mAs corresponds to 13 points, 1 point less would be 12 points. 12 points correspond to 16 mAs. The settings for the exposure are hence 77 kV and 16 mAs.

Recalculating the mAs when changing the film-screen sensitivity class

In the table (Fig. 9.18), 109 kV and 10 mAs with a screen combination of sensitivity 400 are found for a gastrointestinal overview (lying) in an overtable system.

According to the table (Fig. 9.20), a transition to a film-screen system with a higher sensitivity of 800 in a voltage range (U_R) of 90 to 150 kV equates with a decrease of 4 exposure points (from -4 to -8), i.e., that points from the kV and mAs can be reduced by 3. In the table (Fig. 9.19), 10 mAs corresponds to 10 points. Subtracting 3 points yields 7 points, and 7 points correspond to 5 mAs. The settings for the exposure are hence 125 kV and 5 mAs.

A similar result could also have been obtained by making the following estimation: When changing to a film-screen system that is twice as sensitive, the amount of radiation, i.e., the mAs, can be reduced by approximately one-half, i.e., to 5 mAs.

Using the point table for conventional tomography

By using the point table, the exposure data for tomography can be approximated from the normal projection X-ray. First, you need to factor in any deviations from the standard conditions of the exposure table by making appropriate adjustments as portrayed above, i.e., precisely identify the normal imaging data. Then preferably raise the mAs by the following points, such as +2 points for linear slices, +1 point for zonography.

Since the time for the planigraphic movement of the tomograph is preset, this planigraphic time must be set as the imaging time in non-automatic operation. The mA that also needs to be set is calculated from the mAs: mA = mAs/imaging time. The kV, mA and s are hence set (3-point radiography since there is no automatic exposure control; see section 9.1).

Relationship between the exposure points, region of interest and exposure area

The two exposure values that correspond to the bottom and top end point of the straight part of the gradation curve are important for good film illumination (see section 8.2). They mark the area of exposure. The difference between the logarithms of these two exposure values is termed the exposure area. As long as the area of interest, i.e., the radiation contrast of an object, lies within this exposure area, imaging occurs in the straight part of the gradation curve, that is, with optimum contrast (Fig. 9.22, left).

In Fig. 9.22 (top right), the exposure area extends from 1.5 to 2.4 and accordingly equals 2.4 − 1.5 = 0.9. The region of interest is exactly in the middle, i.e., it extends from 1.65 to 2.25 and therefore equals 0.6. The exposure points from the point table are simultaneously entered. As mentioned previously, 3 EPs mean that the exposure, or dose, is doubled. 1 EP more raises the exposure by approximately 25%.

The area of interest in Fig. 9.22 is smaller than the exposure area, and therefore the chosen mAs is relatively uncritical. A somewhat lower kV could

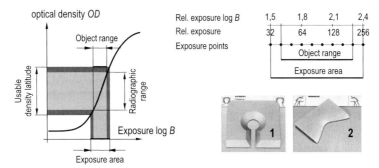

Figure 9.22 Illustration of the area of interest and exposure area (top left and right). Homogenizing filter for insertion in the accessory rails of the collimator: For filters for skull X-rays (child skull (1) in this instance) and shoulder X-rays (2), see chapter 4.

have been chosen (to enlarge the area of interest); even then the area of interest would lie within the exposure area.

The number series 32, 64, 128, etc., in Fig. 9.22, top right, is the relative exposure by the sensitometer (exposure of 21 steps; see section 8.1, Fig. 8.7). The numbers mean that the exposure is 32, 64, 128, etc., times larger than in step 1. Above, the logarithms of the numbers 32, 64, 128 are given, i.e., the logarithm of the relative exposure (for logarithms, see the appendix).

Fig. 9.22 (left) portrays the relationship between the area of interest, exposure area and radiographic range on the density curve. The exposure must be such that the object contrast range lies within the exposure area; only then is the object contrast range shown with optimum contrast.

Fig 9.22 (right) shows that the region of interest would still be within the exposure area in this favorable scenario, even when the exposure is increased or reduced by 1.5 EP. It would of course be better to lower the exposure by 1.5 EP to reduce radiation exposure.

A more contrast-rich image can be obtained by increasing the gradation of the film (different film material, improve the film processing), or by enlarging the region of interest (low kV).

If the object contrast range is greater than the exposure area, the reverse is true. Critical objects in this regard are those that have very high differences in density such as the skull, shoulder, pelvis (especially in standing patients), cervical spine, pelvic spine, foot, etc. Graduated screens and homogenizing filters can help in this instance.

Homogenizing filters and semitransparent diaphragms act in front of the object. Their shape adapted to the object prevents too much object contrast from arising by attenuating the parts of the primary beam cone that contact less-absorbing regions of interest or edge parts of the object. They necessarily harden the radiation in these areas, which makes the images there less contrast-rich. Such special homogenizing filters exist for several particularly critical organs (Fig. 9.22). They can be inserted in the accessory rails of the collimator.

With fluoroscopy and indirect radiography as well, radiation in less absorbing areas or border zones must be avoided at all costs because such radiation (such as too much light in the image intensifier) can worsen the contrast of neighboring areas in the X-ray.

Troubleshooting inferior image quality

If the image quality is not satisfactory or the quality of the X-ray spontaneously changes, it may be for the following reasons:

- Film is too bright:
  - The film or screen is less sensitive than usual
  - The film sensitivity is not appropriate for the screen
  - More strongly absorbing grid than usual
  - Wrong measuring field has been set
  - Larger SID than usual
  - Voltage response of the screen not factored in
  - Screen improperly inserted
  - Developer temperature is too low
  - Blackening correction switch is improperly set
- Film is too dark:
  - The film or screen is more sensitive than usual
  - Less absorbing grid or no grid
  - Wrong measuring field has been set
  - Smaller SID than usual
  - Blackening correction switch is improperly set
  - Developer temperature is too high

Other problems with the image and their probable causes:

- Insufficient film definition: voltage is too low
- Film is faint: voltage is too high
- Film nearly black: grid does not lie in the beam path, or wrong screen
- Low-contrast film with brown tinges: Out of developer? Regeneration pump failed?

- General gray veil in film: Film overlapped, previously irradiated, developer temperature too high or not properly focused in simultaneous alternating operation.
- Film blackened differently from edge to edge: Tube off center, or the grid is out of focus
- Film partially blurred: pressure unevenly applied to cassette, or cassette not correctly closed
- Film is spotty: screen dirty, or film touched with moist fingers
- Edges of film are partially black: cassette not light-tight, film partially pre-exposed, or the cassette not correctly closed
- Spotfilm uniformly unsharp: patient moved, or the wrong focus was selected
- Film shows islands of splash-like bright areas: contrast agent on tabletop or draping
- Film shows surface soiling: film processor rolls are dirty

9.1.7 Icons for medical technology systems

To make a diagnosis, users must understand the projection and geometry options of the different X-ray systems. These are described in detail by the manufacturers of the respective X-ray system in the user manuals and data sheets.

The number of icons and symbols used in X-ray systems is still very large. The digitizing of X-ray systems and subsequent use of computer workstations has increased the number of the icons and symbols. In addition, different manufacturers use and will continue to create different signs and icons. The manufacturer's user manuals should therefore be used as the binding reference for the meaning of icons and symbols. Fig. 9.23 shows a small selection of symbols used on a generator control console.

Meaning of the icons in Fig. 9.23:

1) Workstation selection (of the imaging system)

2) Display of density compensation

3) Display of mode of operation (direct or indirect exposure)

4) Tomography display

5) Selection of automatic exposure control (Iontomat)

6) Setting of blackening compensation (exposure points)

7) Selection of film-screen combination

8) Key for display of exposure data of selected X-ray tube

9a) Display of kV (81 kV in this instance)

9 X-ray Imaging Technology

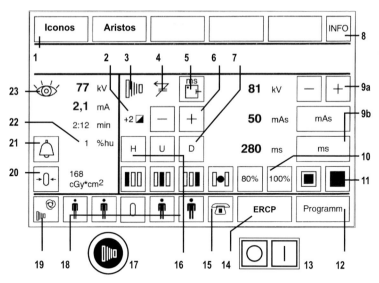

Figure 9.23 Example of an arrangement of control elements and their icons in a generator control panel

9b) Display/redisplay of mAs/ms or selection change
10) Selection of tube load
11) Selection of small/large focus
12) Switch to organ programs
13) Turn on/off X-ray system
14) Display of selected organ program
15) Display for teleservice
16) Selection of film-screen combination
17) Switch to start the exposure
18) Selection of transparency compensation
19) Display of 0-point technique, on/off
20) Reset fluoroscopy data and dose area product
21) Reset fluoroscopy warning signal
22) Display of current tube load
23) Icon for the fluoroscopy data field

Control elements generally come in the form of units for system operation (e.g., positioning the patient and setting the system geometry) and generator operation (e.g., generator data, image parameter selection and image resolution) that are together or separate and/or close to or distant from the patient.

9.1 Radiographic Parameters

Control elements such as keyboards, mice, touchscreens, etc., for image processing and evaluation are integrated in workstations.

syngo

syngo™ is a universal user interface that runs on Windows™ and comes with all Siemens digital imaging systems. The easily recognizable icons make operation intuitive, regardless of the workstation and imaging system (radiography, fluoroscopy, angiography, MR, CT, etc.). Communication within and between hospitals is ensured by the adoption of the DICOM standard.

The same user interface is used for all imaging modalities from the application to postprocessing. The user software supports all standard applications such as patient lists, image previews, interactive image compilation, 3D displays, vascular analyses, cardiac applications and archiving.

A detailed description of all the functions associated with the icons is beyond the scope of this work. Therefore only a few examples that appear on the monitor are shown in Fig. 9.24.

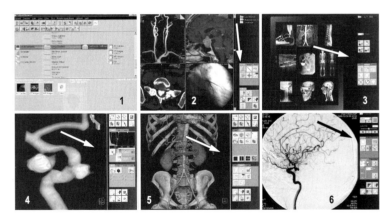

Figure 9.24 The user interface syngo with icons (arrows)

1) Patient list: Access to all patient and examination data
2) Viewer: Preview of the images from the patient list
3) Films: Interactive compilation of sheet film
4) InSpace 3D: Display of three-dimensional views
5) InSpace viewer: For online diagnosis and intervention
6) Angio: DSA postprocessing independent of the imaging system

9.2 Digital Exposure Technique

Digital imaging and processing have long since proven their importance for medical diagnostics and intervention. The technology will continue to provide new opportunities – both with regard to image quality and global communication to the benefit of the healthcare industry.

This chapter addresses the foundations for understanding digital radiography.

The term "digital radiography, DR" refers to all digital exposure methods in projection radiography. All methods use processors to convert the analog radiation pattern resulting at the imaging system into digital data. This digital image data can be processed in different processes: Image display and diagnosis, frame buffering, image processing/postprocessing, image archiving, network connection, etc. They can also be created for special evaluations or archiving via laser cameras as film images.

As already described in chapter 8, the imaging applications established in modern digital radiography (DR) are as follows:

- DFR = Digital Fluoro Radiography (image intensifier radiography or digital fluoroscopy) with or without CCD systems.
  DSA = Digital Subtraction Angiography and
  DCM = Digital Cine Mode, unsubtracted dynamic studies in cardiac angiography are special forms of DFR.
- DLR = Digital Luminescence Radiography (imaging plate systems, also CR = Computed Radiography) digital exposure technique.
- FD = Digital radiography with flat detectors (solid state detectors, flat-panel detectors) for digital radioscopy, fluoroscopy and angiography.

9.2.1 Digitization, digital technique

Digitization refers to the process of converting analog signals into digital ones. The digital technique is the technical procedure of processing and transmitting digitized signals or signals created by digitization.

All digital imaging methods are based on the same digital generation and processing procedure which begins with the digitization of the analog image signals in the analog-digital converter (ADC) and concludes when necessary with the conversion of the digital image signals in the digital-analog converter (DAC) back into an analog image (Fig. 9.25). Digital-analog conversion is necessary for viewing and assessing the moving or static result image on a CRT monitor as well as on modern LCD monitors.

Figure 9.26 schematically shows the process of digitizing analog signals, whereas the values and curves are only symbolic. They were not converted in real. The analog voltage signal, e.g., at the image intensifier output (video

9.2 Digital Exposure Technique

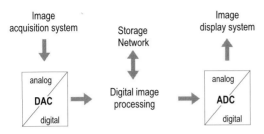

Figure 9.25 The basic process from an analog to a digital image

signal), including its different frequencies (a) is scanned line-by-line (1) as a separate value or via single signal samples (b) at a certain time interval t_1 and, e.g., using the I.I. exposure technique. These sequences (2) are available as defined voltage values (c) for conversion into a binary code.

Defined voltage values are already read out (c) for digital imaging systems, such as imaging plates (CR), CCD sensors and flat detectors (also see chapter 8).

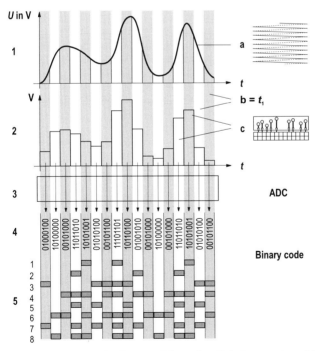

Figure 9.26 Schematic diagram of the analog-digital conversion of the sequence of image information, voltage values U in V of the image signals. In the case of image intensifier radiography, as a function of time t.

9 X-ray Imaging Technology

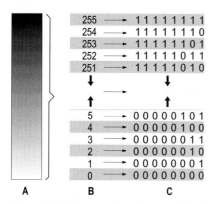

Figure 9.27 Binary code and memory depth or contrast resolution

The defined voltage values U which each correspond to a particular density value of the image are assigned in the analog-digital converter (Fig. 9.26-3) to the corresponding binary numbers (Fig. 9.26-4) (binary code). The average intensity values of the pixels (Fig. 9.26-c) are also called the average density values or local grayscale values. As the number of figures in the binary code increases, the definition of the gray level of the individual pieces of image information, of the pixel, becomes finer. The schematic diagram in Figure 9.26 is based on the example of a grayscale value range of 256 grayscale levels (0 to 255), which corresponds to an 8-bit "memory depth" (contrast resolution) of 1 to 8 (also see Figure 9.27).

Sampling theorem

When preparing the analog image signal at the image intensifier output, the sampling frequency t_1 must be at least less than half of the frequency occurring in the analog image for error-free reconstruction according to the sampling theorem.

Example: If the highest frequency occurring in an analog image is approx. 20 to 22 MHz, the sampling frequency of the analog-digital converter (ADC) must be 45 MHz. The individual time interval of the signal samples is then $t_1 = 22$ ns.

Binary code and memory depth

Computer, as used not only in digital medical imaging, perform the computing operations according to the binary or dual number system (binary digit 0 or 1; dual, two possibilities).

In the analog-digital converter (Fig. 9.26-3, ADC), every analog value of a single piece of image information (density value) which is available in the form of a specific voltage value is assigned to a binary code which is comprised of a series of ones (1) and zeros (0):

Figure 9.27 uses the example of 256 gray levels to show:

- A) The absolute grayscale range from black to white.
- B) The division of this grayscale range into 256 individual levels. Every 256th segment has a defined/average intensity or grayscale value.
- C) Each of these 256 gray levels is assigned one 8-figure binary number comprised of 8 bits (8-bit depth).

The bit information unit

The information unit, the bit, short for binary digit, is the term for the smallest representation unit for numbers in binary number representation. It is a pulse with the unit "bit" which logically represents a 0 or 1. Logical "zeros" or "ones" are two different low voltage pulses. During computing with the digital numbers in a computer, the smaller voltage value is assigned to 0 and the larger voltage value is assigned to 1.

Or: The information unit 1 or 0 corresponds to an electronic control state, which may also be referred to as "on" or "off" or potential 1 and potential 0.

The bit indicates the number of possibilities for defining a grayscale value (from black to white). For example, the term "8-bit depth" (also memory depth) is used. 8 is the number of exponents, i.e., an 8-bit depth corresponds to $2^8 = 256$, that means 256 possibilities of a grayscale value from black to white for the grayscale definition of a pixel (an image element).

Figure 9.28 shows: A) An image with a grayscale value of 1 bit (depth) would have only $2^1 = 2$ possibilities to define this grayscale value, i.e., there would only be black-white image portions for defining the grayscale of the pixel. B) An image with 2 bits (depth) = $2^2 = 4$ gray levels. C) An image with 3 bits (depth) = $2^3 = 8$ gray levels, etc. Today, 8 or 10 bits are used in digital

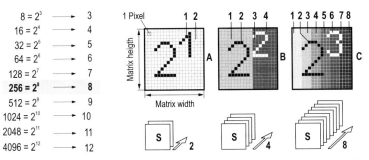

Figure 9.28 Schematic diagram of the grayscale values in bits (grayscale resolution)

radiology, and primarily a 12-bit depth is used in angiography, i.e., the possibility of 4,096 different grayscale levels.

The storage space is made available in the frame buffer for every pixel for the number (binary number) that represents its grayscale value. The term memory depth indicates that, for example, a 10-bit memory must be "deeper" than an 8-bit memory, i.e., must contain more memory locations (Fig. 9.28-S).

Matrix

In the case of I.I. TV imaging with linear scanning of the image from the left to the right and from the top to the bottom as well as the horizontal division of the lines (signal samples t_1), an image divided into equal partial areas is produced. These partial areas or image elements are called pixels (picture elements).

In the case of imaging systems (CCD sensors, imaging plates, flat detectors), the image information is already in the form of pixels. The reading out of pixel/image information at CCD sensors, imaging plates and flat detector systems was already described in chapter 8.

The matrix indicates the number of pixels in the horizontal direction times the number of pixels in the vertical direction. The total number of pixels, i.e., the number of pixels, available for visualizing the diagnostic result image can be calculated from this. *Example:* 512 pixels in the horizontal direction × 512 pixels in the vertical direction yield $512^2 = 262, 144$ pixels. In the case of a matrix $1,024 \times 1,024$ ($1,024^2$), 1,048,576 pixels are available for visualizing the image. Since the number of pixels is typically equal in the horizontal and vertical direction, 512^2 is often written as 0.5k matrix and $1,024^2$ as 1k matrix.

Figure 9.29 schematically shows a matrix 512^2 (A), i.e., 512 pixels in the horizontal direction and 512 pixels in the vertical direction, and a matrix $1,024^2$ (B), i.e., 1,024 pixels in the horizontal direction and 1,024 pixels in

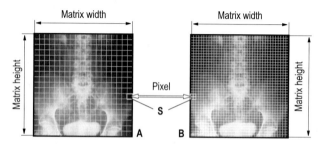

Figure 9.29 Schematic diagram of a matrix

the vertical direction. The gray level information (S) is 4 times greater for the same memory depth (bit depth).

- As the image matrix increases so does the precision of the geometric resolution, i.e., smaller details can be visualized.
- As the memory depth (bit) increases, the grayscale value gradation becomes finer and smaller absorption differences can be visualized, i.e., greater diagnostic differentiations can be made.

Bytes and memory capacity

Therefore, an image with a grayscale value of 8 bits (depth) has $2^8 = 256$ possibilities, and as a result, an image with 12 bits has $2^{12} = 4,096$ possibilities for grayscale resolution in a pixel.

In practice when indicating the pixel quantity in connection with the grayscale resolution with which an image is displayed, reference is made, for example, to a 512^2/8-bit matrix or $1,024^2$/12-bit or 0.5k matrix/8 or 1k matrix/12.

To quantify the storage space, a byte is defined as

- 1 byte = 8 bits.

The necessary memory capacity (storage space) depends on

- The matrix size and
- The bit depth

Example: 512^2/8-bit matrix:

An image with an 8-bit depth and a 512^2 matrix has 262,144 pixels × 1 byte (8 bits), i.e., requires a storage space of $512 \times 512 \times 1 = 262,144$ bytes.

Example: 1.0242/8-bit matrix:

An image with an 8-bit depth and a $1,024^2$ matrix requires 4 times the storage space, $1,024 \times 1,024 \times 1 = 1,048,576$ bytes.

Example: 1.024^2/12-bit matrix:

An image with an 12-bit depth and a $1,024^2$ matrix requires 6 times the storage space, $1,024 \times 1,024 \times 1.5 = 1,572,864$ bytes.

- 1 kb (kilobyte)   =   1,000 bytes
- 1 MB (megabyte) =   1,000,000 bytes
- 1 GB (gigabyte)   = 1,000,000,000 bytes

The data is raw image data that is only rarely stored. Already processed images containing significantly less image data are typically stored. It is also possible to reduce or compress the image data. As a result, a created (acquired) $1,024^2$-matrix image can be compressed to a 512^2 matrix.

Data compression, data reduction

Data compression is a compression of the data stream (the data quantity when transferring in bits per second) by removing bits on the transmission side. In this process, the receiver side (e.g., the storage medium) receives information as to how and where the bits are to be used so that the data stream can flow back without restrictions.

During data reduction, non-essential bits are removed from the data stream. This data is no longer available for the reconstruction of the image. This is data that is virtually cached so that it does not lose any diagnostic information content.

Storing and archiving

Like every digital data processing system, the digital imaging system of the diagnostic X-ray system also has a working memory. This is the RAM, random access memory. It is the memory with open access, the working or main memory of a computer system.

When booting the computer, the operating system and the user programs are loaded from the hard disk into the RAM, i.e., the corresponding data is loaded from the permanently installed hard disk into the working memory. The system now communicates without additional commands only with the RAM.

When switching off the system, all data in the RAM is lost. Therefore, all desired information (image and patient data, etc.) must be previously transferred to the hard disk (Winchester disk) with the appropriate save command.

Winchester disk

The Winchester disk is a permanently installed storage medium in which one or more magnetic disks are arranged over one another on a spindle. The data can be deleted or overwritten on the Winchester disk. It is provided for the relatively short-term storage of data.

CD drive

Data can also be transferred from the RAM (from the hard disk or Winchester disk) to the exchangeable media, e.g., to optical disks (OD) or magneto-optical disks (MOD). The advantage of exchangeable media is offline information exchange. ODs are data carriers that can only be written to once. Data on MODs can be deleted or overwritten. Both storage mediums provide substantial memory capacity for several thousand images, increasing tendency. They are relatively robust and have a service life of over 40 years.

Server

For online information exchange, the data can also be transferred from the RAM into the server. Servers, also called file-server computers, are central memories connected to a network which other interested parties (departments, clients) can access. This can occur locally in a hospital, centrally for a hospital network, or globally (see chapter 10, Patient data management).

Online, offline

Online is the term for a direct connection (e.g., via cable lines) of two or more modalities so that direct, virtually delay-free communication can occur between the modalities. During offline information exchange, there is no such direct connection between the modalities.

Online and offline refer to the time difference between the creation of the data and their processing. During online operation, data is processed immediately in real-time. When offline, any length time period can be allowed between the inputting and processing of data.

9.2.2 Digital radiography operating types

The digitization of a diagnostic exposure or exposure series allows the image to be viewed almost immediately, in real-time (live image) on the monitor (exception: imaging plates).

Different computer-aided parameters (also software tools) are available for generating a digitized image. As operating types, they determine the requirements for targeted application demands that optimally support diagnosis.

Last Image Hold (LIH)

After radiation is switched off, the last fluoroscopic image is retained in a local cache and consequently on the monitor. This is referred to as the last image hold (LIH).

Single image and series exposures

Digital radiography systems without a fluoroscopy device (without fluoroscopy) principally have only one operating mode, i.e., that of the digital single exposure.

Digital fluoroscopy systems for universal and internal fluoroscopy, for special vessel and cardiac angiography, or for interventional procedures can generate diagnostically relevant exposures in the form of single images or image series during the continuous fluoroscopy.

Series exposures are possible with a time-controlled, automatic sequence of image exposure rate (s) or manually with variable image rates, e.g., of 0.5 to 7.5 frames per second (f/s).

Fluoro loop, frame grabbing

Fluoroscopy sequences can be stored via fluoro loop for subsequent use.

Video loop

Video loop is the digital image acquisition of fluoroscopy series (with a minimal fluoroscopy dose) at, e.g., 15 or 30 frames per second. Dynamic fluoroscopy sequences (video loop) can be retrieved at any time from the local cache and displayed on the monitor.

Digital pulsed fluoroscopy

For dose reduction, especially in time-intensive fluoroscopic examinations, it is possible with the image intensifier technique to create only one image at half the dose rate by integrating two image periods.

Figure 9.30 shows the operating type of continuous fluoroscopy:

- 1: At 100% dose (A1). Image reconstruction and reading out (scanning) of image information (B1). Display on the monitor of one scanned image period (C1). This image reconstruction also corresponds to the analog I.I. TV technology.

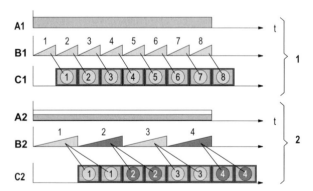

Figure 9.30 Schematic diagram of the time sequence in 1: Standard fluoroscopy (continuous) and 2: Fluoroscopy with half of the dose rate (SUPERVISION).

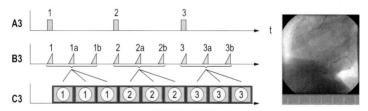

Figure 9.31 Schematic diagram of the time sequence during pulsed fluoroscopy (CAREVISION) and an ECG-triggered exposure (right)

- 2: Continuous fluoroscopy at 50% dose (A2). Image reconstruction and reading out of image information (B2). Display on the monitor over two image periods (C2). This image reconstruction also corresponds to the analog I.I. TV technology. However, the halving of the dose rate (2) results in greater noise in the image which is then reduced by a special image processing mode, the digital averaging of a plurality of individual images.

Pulsed fluoroscopy represents a further measure for reducing the dose in digital image intensifier or flat detector fluoroscopy.

Figure 9.31 shows radiation pulses at 100% dose (A3) and image reconstruction (B3). To display a constant image on the monitor, the stored radiation image (1, 2, 3, etc.) is filled in by multiple repetitions in the pauses between the radiation pulses (1a, 1b, etc.). This procedure is referred to as intermediate image buffering or gap filling (B3). In this context, depending on the pulse frequency (pulse pause), a certain strobe effect occurs more or less since movement of the object to be examined cannot be ruled out in the pulse pauses or, for example, the catheter has been further inserted. Display on the monitor via image integration (C3).

In contrast to continuous fluoroscopy, the X-ray radiation is interrupted at equal time intervals during pulsed fluoroscopy. Digital pulsed fluoroscopy can be performed, for example, with 15 or 30 p/s (pulses per second) (in pediatrics up to 60 p/s). The pulse frequencies can be adapted to the particular application requirements (exposure modes) for a significant reduction in the radiation exposure which is particularly advantageous in interventional procedures.

Pulsed fluoroscopy can be fundamentally achieved with two alternative technical solutions:

- Via a grid-controlled tube, secondary pulsed fluoroscopy, i.e., electron emission in the tube, is interrupted at regular time intervals (see chapter 7).
- Via the primary pulsing generator, i.e., the X-ray radiation is interrupted at regular time intervals by the generator.

The grid-controlled X-ray tube is used primarily for cardiac angiography and pediatric examinations.

In the case of technical data, pulse frequency information is often supplemented by the note that pulsed fluoroscopy is performed with "digital real-time filtering" and "slidingly weighted averaging with a motion detector."

Real-time filtering

In the case of pulsed digital fluoroscopy or digital image series, the noise, motion artifacts and radiation dose are reduced via time filtering.

Time or recursive filtering is the integration of pixels of a memory output image into the memory input image. Recursive means returning to known values and that comparable pixels of an image series are added in the computer by repeating this procedure. Since this computing procedure occurs according to a certain repeating scheme, i.e., a direct algorithmic process, this is also referred to as digital real-time filtering.

Arithmetic averaging

To keep the exposure to radiation minimal during fluoroscopy, angiography and interventions, the dose per image is set to be as low as possible. An increased noise range is quickly reached in this process (quantum noise, see section 9.3). Noise reduction results in an improvement by integrating and averaging a number of image contents (e.g., 4, 8, 16 or 32). The disturbing quantum noise is statistically distributed and suppressed with respect to the image signal.

When averaging is performed arithmetically for noise reduction, the last four images are added up and then averaged by dividing by four, for example. The thus calculated image equally contains 1/4 of each of the last four images.

Slidingly weighted averaging

Slidingly weighted averaging is another type of averaging which reduces the negative influence of organ movement in that less recent images are factored less into the image shown on the monitor than more recent ones. The percentage of less recent images to the current image decreases with every newly received image (1/2, 1/4, 1/8, 1/16, etc.). The average is not arithmetic but differentiated (nuanced), i.e., "slidingly weighted." This computing procedure is also referred to as recursive averaging.

Motion detector

When visualizing moving objects, the time smear effect, i.e., a smearing of the pixel information due to object movement, can occur during every image integration (time or recursive filtering).

A motion detection recognizes the pixel values significantly varying at these points and controls the integration of pixels into the memory input image.

This computing process is referred to as slidingly weighted averaging with a motion detector.

Digital subtraction angiography (DSA)

Digital subtraction angiography is an X-ray diagnostic method for the isolated visualization of vessels.

Figure 9.32 shows the process of a DSA and the time course of the contrast agent concentration in the target area: Precontrast exposures (pre-injection exposures) are first created and are combined to form one image, mask (M) and stored before the contrast agent enters the target area. During the enrichment of the vessel with contrast agent in the diagnostically relevant region, opaque radiographs (F) are recorded and also stored with selected image exposure rates. Precontrast and opaque radiographs are subtracted from one another in the image processor and displayed on the monitor in real-time as subtraction image (S). The comprehensive image processing possibilities allow the result image to be optimized for diagnosis (see section 9.2.3).

DSA is generally performed with an image matrix of $1,024^2$ (in some instances also with 512^2) and a grayscale resolution (depth) of 10 to 12 bits.

The two possibilities of continuous or pulsed image acquisition are available.

- During continuous operation
higher image frequencies of 15 and 30 frames per second, for example, are possible. It is preferably used in cardiac angiography or pediatric examinations (in pediatric examinations also 60 frames per second). During continuous operation, an integration (summation) of a number of precontrast

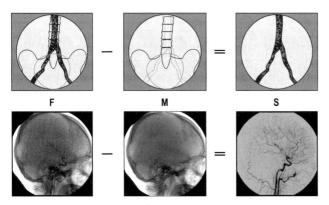

Figure 9.32 Digital subtraction angiography (DSA): Principle based on the example of an aortography and a result image of the carotid communis.

9 X-ray Imaging Technology

as well as opaque radiographs for improving the contrast resolution and the signal-to-noise ratio is performed.
- During pulsed operation
image frequencies of 0.5 to 7.5 frames per second, for example, are typical. In contrast to continuous operation, the relatively high dose (good quantum yield) of the individual short pulses has the advantage of improved contrast resolution or more favorable signal-to-noise distance. Both the patient exposure and the tube stress are reduced. Image integration as during continuous operation is also possible during pulsed operation.

Roadmapping

Roadmapping is subtracted fluoroscopy in order to be able to observe and manipulate the position of catheters, guide wires or coils in the vessel during a fluoroscopic examination. Overlapping by bones or material for embolization, for example, is eliminated.

Figure 9.33 shows the basic roadmapping procedure: In the roadmapping operation type, fluoroscopy automatically switches to the subtraction mode (1). During fluoroscopy a small amount of contrast agent is injected (2). If the vessel is filled with contrast agent, the examining physician interrupts the fluoroscopy procedure. The computer inverts the last opaque radiograph (4) and stores it as a mask (4 and 8). After fluoroscopy is switched back on, the subsequent fluoroscopic images are subtracted from the mask. The wire or catheter is shifted by the examining physician under observation on the vessel image (5 through 7).

Overlay fade and dynamap

Overlay fade is an online superimposition of the active fluoroscopic image and a reference image. This prevents motion artifacts from occurring during

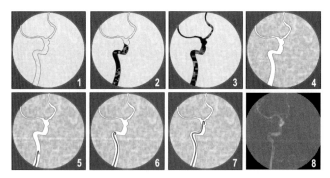

Figure 9.33 Roadmapping. Principle based on the example of an aortography; right bottom: Subtracted and inverted mask image.

230

roadmapping, for example. In the case of a static display of the superimposed images, the term overlay is used, and dynamap is used in the case of dynamic scenes.

Pre- and post-compare

Pre- and post-compare is a function for comparing dynamic scenes. The vessel situations can be compared before, during, and following the intervention and, for example, two scenes can be displayed at the same time on the monitor.

ECG triggering

Subtraction artifacts that may occur due to inexact coverage during subtraction can be avoided via a heart phase-controlled exposure operation (ECG triggering). Heart phase-controlled exposure operation is possible in continuous as well as in pulsed operation.

Digital cine mode (DCM)

To visualize dynamic heart function studies, the bright dynamic image of the I.I. output screen was filmed with a cine camera in the early days of digital radiography. 35-mm cine cameras (roll film cameras) with an image frequency of up to 60 images per second were used in cardiac angiography. This technique was replaced by digital exposure sequences with the function of the digital cine mode CDM.

Peripheral digital angiography

For exposures in dynamic angiography, there are different technical processes for achieving peripheral digital angiography. An image series of the arteries from the pelvis region to the feet is fundamentally made during combined arteriography of the abdomen and the lower extremities.

The exposures of the individual image areas are made either by shifting the patient table, i.e., with the patient, or are made by the imaging system, i.e., with a stationary patient.

In addition to the method of visualizing every vessel segment with a separate contrast agent injection, there are different automated exposure procedures for peripheral stepwise shifting:

- Continuous with digital subtraction (Perivision^{TM})
- Continuous with digital imaging without subtraction (Peristepping^{TM})
- Variable with digital imaging (Periscanning^{TM})

Perivision (Fig. 9.34) is peripheral stepwise shifting with digital subtraction with only one contrast agent injection (DSA technique).

9 X-ray Imaging Technology

A: The imaging system starts at a position (S). The collimation is defined and stored for every individual exposure position (1a through 1f). The imaging system is now in a position (E).

B: The examination begins with the acquisition of masks for every individual segment from the feet to the pelvis (2f through 2a). The steps are switched automatically. The imaging system is now in position (S).

C: After acquisition of the last mask image, the contrast agent injection is triggered. The filling procedure starts with preselected image frequencies (3a through 3f) and the stepwise shift is switched manually. The imaging system is now back in position (E).

D: The subtracted images are already available during the exposure series (online subtraction visualization).

Peristepping (Fig. 9.34) is peripheral stepwise shifting in digital imaging without subtraction and with only one contrast agent injection.

A: The imaging system starts at a position (S). The collimation is defined and stored for every individual exposure position (1a through 1f) prior to the start of the exposure series. The imaging system is now in position (E). The imaging system is then returned to position (S).

Contrast agent injection is triggered prior to the start of the exposure series. The filling course goes from the abdomen to the feet with preselected variable image frequencies (3a through 3f), and the stepwise shifting to the particular exposure positions is switched manually under observation of the contrast agent bolus.

Periscanning (bolus chase) is variable peripheral shifting in digital imaging without subtraction and with only one contrast agent injection.

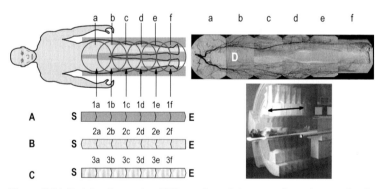

Figure 9.34 Peripheral stepwise shifting and result image as the perimap or longleg visualization of the complete peripheral vascular system

In contrast to "classic" visualization with fixed steps as in perivision and peristepping, the distances of the imaging positions are variable in this exposure technique. The imaging system or the patient table are moved continuously under observation of the contrast agent bolus. The individual half images are each manually triggered at the instant of the best possible contrast agent filling.

This variation of the dynamic peripheral angiography typically requires 50-60 exposures. The images are subsequently assembled so that the entire vessel region is optimally displayed.

Rotation angiography

A progressive variation in dynamic angiography is the exposure procedure involving rotation of the imaging system in a circular arc around the patient known as rotation angiography (Dynavision^{TM}, Fig. 9.35). Rotation angiography can be used to spatially visualize two-dimensional images, e.g., complex vessel images, via special exposure modes and computing operations as 3D images (three-dimensional).

The exposure series can be made under 3D-defined exposure conditions for all exposure systems with orbital rotation processes (Fig. 9.35, right) and transferred to a special workstation. A 3D data set with all current digital image processing functions is calculated there from the projection exposures in the system processor.

Three-dimensional visualizations (3D reconstructions) support the visualization of complex structures or overlapping vessels (with only one contrast agent injection) and allow significantly improved definition of the individual steps for upcoming interventions.

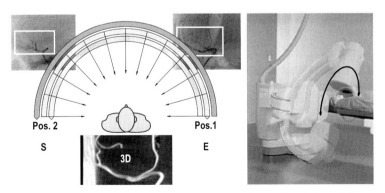

Figure 9.35 Rotation angiography (Dynavision), examples of two 2D result images from the LAO/RAO projection and the 3D image calculated from this

With respect to Fig. 9.35: The angle range between the individual exposures is defined (from pos. 1 to pos. 2) prior to starting the exposure series. The imaging system is now in starting position (S).

The examination begins with the creation of a mask (precontrast image) for every previously defined angle position (from pos. 2 to pos. 1). The imaging system is now in starting position (E). Contrast agent injection is triggered prior to the start of the acquisition of the opaque radiographs.

With a rotation speed of up to 40°/s and an angle range of up to 200°, the exposures are triggered according to the entered angles. Up to 80 image projections are possible in this context (from pos. 1 to pos. 2). A special SW program calculates a 3D data set (3D) from the projection exposures.

The "angle triggering" exposure procedure renders possible a dose savings via the image frequency reduction as well as a significant time savings. If a pixel shift is necessary, this can be performed automatically in several consecutive image pairs.

Rotation angiography under 3D-defined exposure conditions (3D reconstruction) requires the object to be in the isocenter of a C-arm exposure system (see chapter 9 – isocenter).

The angle position of the imaging system can be automatically assumed according to the selected reference image to reconstruct the exposure data.

3D imaging in surgery and computed tomography

Movable C-arm systems for surgery and computed tomography (chapter 6) are among the radiography systems with an isocenter. Rotation fluoroscopy with the possibility for 3D reconstruction can also be performed with these radiography systems comparable to the previously described angiography systems.

In the case of surgical C-arms, a fixed number of fluoroscopic images can be acquired at fixed angle distances given an orbital angle range of 190° from an object and an automated continuous orbital rotation.

In the case of computed tomography, the object is always in the isocenter as a result of the rotating imaging method and the consequently derived constructive system structure. See the appendix for specialized literature on computed tomography.

9.2.3 Image processing in digital radiography

The use of detectors as image recorders first made digital image processing possible in computed tomography. The introduction of the digitization of analog image signals at the image intensifier output marked the start of the constant development of digital image processing.

9.2 Digital Exposure Technique

Regardless of the digital imaging system technology (chapter 8), image postprocessing principally provides the same parameters, i.e., procedures, for all digitally stored images.

Radiography systems with digital imaging provide a number of possibilities to make changes to the digitally stored pixel values of a raw data image via mathematical methods (algorithms). The quality of the result image still depends on the quality of the original image.

Digital image processing cannot add any new information to the original image but the optimal setting for diagnosis can be determined and displayed.

The parameters (procedures) of image postprocessing or image manipulation for improved visualization and assessment include:

- Nat/sub: Switching between precontrast (unsubtracted representation) and subtraction (subtracted representation).
- Invert: Black-white inversion.
- Landmark: Addition of an anatomic background.
- Zoom/magnification: Enlargement.
- Pixel shift: Compensation for motion artifacts.
- Move mask, replace mask (dynamic and static).
- Windowing: Image contrast enhancement.
- Edge enhancement: Special contrast enhancement.
- Max./min. opac.: Maximum/minimum contrast visualization.
- Density optimization: Contrast harmonization.

Nat/Sub, invert, landmark

Figure 9.36 shows the functions:

A: Nat/…switching:
Precontrast representation, unsubtracted representation with contrast.

B:..../sub switching:
Subtracted representation (typical DSA representation).

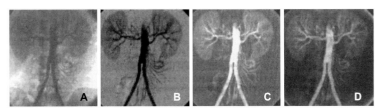

Figure 9.36
Functions nat/sub, invert and landmark based on the example of the abdominal aorta

C: Invert:
Black-white inversion; subtracted and inverted representation.

D: Landmark:
A subtracted image is superimposed by a precontrast image (nat) and the anatomical background is displayed.

Zoom/magnification

Sections (regions of interest) of images displayed on the monitor can be magnified via the zoom or magnification function. Figure 9.37 shows the two variations for magnifying the display of a region of interest. The desired section can be set via image shifting (roam). The magnified section (2x zoom here) is always displayed from the middle of the segment (both images left). The magnification function can be used to magnify any section of a static image (virtually floating) (middle).

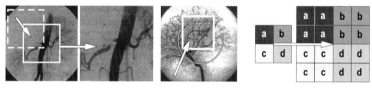

Figure 9.37 Zoom function based on the example of abdominal arteries and the magnification function based on the example of the internal carotid and a schematic diagram of the "pixel magnification"

Figure 9.37 on the right shows the graphic representation of the creation of 2x zoom: If the grayscale value of each individual pixel is assigned to several pixels, an enlarged image segment results, e.g., the a pixel becomes 4 a pixels. This would be 9 pixels for 3x zoom. In contrast to the original pixel, the magnification does not contain any additional information. However, in magnified image sections, image details can often be better recognized.

Dynamic zoom means that scrolling is possible in magnified series exposures.

Pixel shift

Patient reactions to the contrast agent injection can result in motion unsharpness between the precontrast and opaque radiograph, thereby yielding motion artifacts. These can be compensated for via pixel shift. Since pixel shift makes structure noise more visible on the monitor, an attempt should always first be made to change the mask.

9.2 Digital Exposure Technique

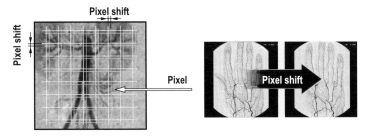

Figure 9.38 Schematic diagram of the pixel shift function

The pixel shift function can be used to shift the mask and opaque radiograph with respect to one another so that both images have the same coverage to the greatest extent possible (Fig. 9.38).

Manual pixel shift:
The mask image is manually shifted vertically and/or horizontally.

Automatic pixel shift:
The region of interest (ROI) is marked in the image. The correct position of the mask image is calculated and automatically shifted within this ROI via an algorithm (Fig. 9.38, right).

Remask

The selection of a mask image that differs with respect to time from the opaque radiograph or vice versa may make improved coverage congruency possible. This motion unsharpness can typically be eliminated by selecting another mask image (remask). Two functions are available for this purpose.

Move mask is the function for selecting another mask for a selected opaque radiograph under observation.

Replace mask is the function for selecting a certain mask image, e.g., to place the mask virtually "blind" exactly in front of the opaque radiograph.

Windowing

Windowing continues to be one of the most important digital image processing methods. The limitation of an available contrast range and "expansion" of this range enhances details with low contrast significantly more clearly.

An image with 10 bits, for example, has a grayscale range (depth) of 1,024 levels. An intensity range of interest and of any size is selected within this entire grayscale range (Fig. 9.39-A). The intensity range of the gray levels (B) to be shown on the monitor is referred to as the window width. The intensity range is changed accordingly by shifting (C) the region of interest (window position). The window width, i.e., the selected grayscale range, is

9 X-ray Imaging Technology

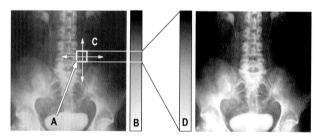

Figure 9.39 Schematic diagram of the windowing function for contrast enhancement

spread over the entire dynamic range of the monitor (D). This intensifies the grayscale range of the image in which the clinically relevant information is contained.

Edge enhancement, spatial frequency filtering

Edge enhancement or image intensity transformation is a method of the spatial filter technique for improving the delimitation of grayscale value transitions. The goal of edge enhancement is to perform contrast enhancement, especially of fine image structures.

The edge enhancement function intensifies transitions from vessels to tissue, for example, so that a multiplication factor is applied to every pixel value located in this transition region from vessel to tissue. The output value of a pixel depends on its input value. Tables known as look-up tables are used to convert the pixel values.

Spatial frequency refers to the structures (contrasts) of an X-ray image within certain regions. Spatial frequency filtering uses algorithms via which the visualized spatial frequency is influenced or changed, i.e., a higher spatial resolution (also see section 9.3.2).

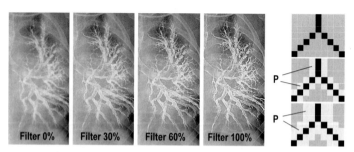

Figure 9.40 Edge enhancement with spatial frequency filtering based on the example of a bronchography with different enhancement factors (filters)

During spatial frequency filtering, the pixel values surrounding the region of interest are also taken into consideration. For example, 3 × 3 = 9 pixels are selected within a quadratic region. This region is referred to as the kernel or filter kernel and is indicated in pixels or millimeters as the edge length. Or: The "vessel edges" are intensified in that a brighter pixel superimposes every pixel that does not differ by a defined value from its neighboring pixel (Fig. 9.40-P).

Taking the logarithm of image signals

The video signal of the TV camera is linearly intensified and digitized for general digital radiography. It must be logarithmically intensified for digital angiography and subtraction angiography.

As a rule, vessels are covered by skeleton and soft tissue structures. The intensity profiles of the image signals therefore differ according to the absorption of the irradiated region.

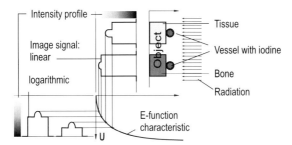

Figure 9.41 As a result of taking the logarithm of the image signal prior to subtraction, vessels can be seen equally well regardless of whether they run over bones or soft tissues

Taking the logarithm of the image signal values compensates for the exponential function of the attenuation law, i.e., the intensity of the vessel image signal is independent of the absorption of the overlapping body layers. For information on the logarithm, see the Appendix.

Look-up tables LUT

The grayscale values of digital images are binary numbers. It is possible to assign these numbers to others according to predefined rules. These rules are defined in the look-up table (LUT).

Figure 9.42 uses the example of a contrast enhancement to show how this influencing of the transmission characteristic of the amplitudes can be used practically. Look-up tables are also used to adapt the signals to the reproduction characteristics of monitors.

9 X-ray Imaging Technology

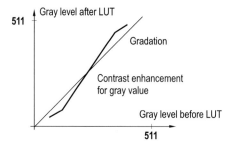

Figure 9.42 Contrast enhancement using look-up tables

For low input grayscale values, the function is used to assign look-up table output values that are less than the input signals. This is followed by the middle range in which LUT output values that yield a steeper curve are assigned. Contrast enhancement is not performed here. LUT output values yielding a flatter curve are again assigned in the top grayscale range.

The contrast of an X-ray image is also determined by the gradation of the film. Two films with different gradation curves produce different images under the same exposure conditions (object and exposure data). Different image impressions can be achieved in approximately comparable manners using digital methods.

Summation/image integration

The image noise (signal-to-noise ratio) is greater for a low dose rate. The increased image noise is decreased via the weighted addition of a number of individual images, i.e., via summation or integration. As the total number of images increases, the signal-to-noise ratio improves (the dose of the individual images basically forms the sum for one result image).

Given minimal injected contrast agent volumes, the vessel filling which would otherwise only be seen over a number of images is visualized in one image. However, flow situations on different sides can be suppressed in this process. During image integration (Fig. 9.43), a number of images are added up (e.g., 1 through 4) for the result shown (5a). The noise and in some cases also the contrast visualization are reduced.

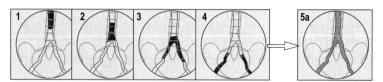

Figure 9.43 Image integration

Max./min. opac.

The max./min. opac (vascular tracing) function stands for maximum peak opacification (with iodine contrast, positive contrast) and minimum peak opacification (with CO_2 contrast, negative contrast) and provides the opportunity particularly in interventional examinations for maximum contrast visualization in one image (See section 9.3 – contrast).

In max./min. opac, only the pixels are combined with the maximum/minimum contrast of the individual images to form one artificial image. The individual vessel sections are acquired and added pixel-by-pixel over the entire image.

Dynamic density optimization

Dynamic density optimization (DDO) refers to the reduction of the dynamic range to emphasize detail contrasts via windowing. Over and underexposed image regions with significant black-white sections are harmonized by the dynamic density optimization function in grayscales (Fig. 9.44).

Roaming

The roaming (image shifting) function allows the image information currently outside of the region of interest to be shifted to the center of the monitor.

Auto-window

The auto-window function can be used to optimally regulate the brightness and the contrast of the image displayed on the monitor.

Auto-shutter

The auto-shutter function allows disturbing, bright image sections to be covered. The image is collimated on the monitor electronically in last image hold.

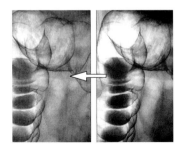
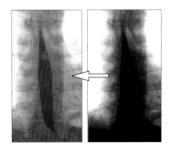

Figure 9.44
Result of dynamic density optimization (DDO): right without and left with DDO

Reference image display

A stored reference image can be displayed on a separate second monitor (Fig. 9.45-1). If only one monitor is available for the fluoroscopic image (live image) and the stored reference image, the following viewing modes are possible:

- Superimposition of the two images (also see overlay fade).
- Next to one another (split screen, vertical).
- Over one another (split screen, horizontal).
- Consecutive, i.e., the live image during fluoroscopy and the selected reference image during fluoroscopy pauses.

Multimap

Multimap is the function for simultaneously displaying up to 16 images on one monitor for visual scene or reference image selection (Fig. 9.45-2).

Echomap

The echomap function can be used for positioning a smaller ultrasound image in the fluoroscopy or radiography image, e.g., in a corner of the monitor (Fig. 9.45-3).

Endomap

Endomap is a software program for digital operation planning and documentation, e.g., for hip prostheses (Fig. 9.45-4).

Automap

The automap function can be used to move the radiographic system (C-arm) into the exposure position displayed by the reference image (reconstruction of the exposure angle). During angulation of the C-arm and given the availability of already stored reference images, these are automatically displayed when using the bi-directional automap function.

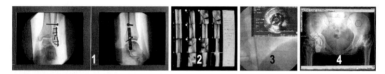

Figure 9.45
Reference image display, multimap, echomap and endomap representations

9.2 Digital Exposure Technique

Graphic collimation without X-ray radiation

The position of the collimators or the semi-transparent filters can be displayed in the last image hold (LIH). Manipulation of the collimators is displayed on the monitor as graphic lines without requiring radiation (Fig. 9.46-A: Careprofile).

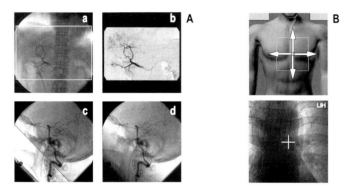

Figure 9.46 Careprofile (A): Graphic collimation on the LIH without fluoroscopy (a) and directly resulting collimator setting for the exposure (b). Placing of the semi-transparent DSA filter on the LIH without fluoroscopy (c) and positioned filter in the fluoroscopic image (d). Careposition (B): Shifting of a virtual, graphically displayed X-ray field.

Shifting a virtual, graphically displayed X-ray field on the live monitor allows the patient to be positioned without radiation (Fig. 9.46-B: Careposition).

Quantitative measurement methods

The possibility of also evaluating digital images quantitatively, e.g., during vessel analysis with stenosis degree determination and distance measurement or a left ventricle analysis for assessing the functioning of the left ventricle, is used in vascular angiography and cardiac angiography in particular.

Quantification has always played an important role in heart diagnostics. In the case of modern digital images, appropriate evaluation programs are used to quickly and precisely perform and log the at times complex and time-intensive calculations for cardiological quantification methods (Fig. 9.47).

Other evaluation functions, e.g., in bone diagnostics, are the calculation of the scoliosis angle (lateral bend of the spine with rotation of the individual vertebral bodies), the spinal displacement or the pelvic tip.

9 X-ray Imaging Technology

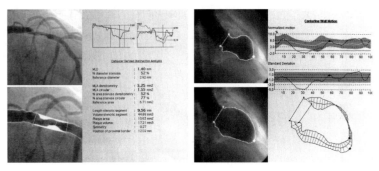

Figure 9.47 Quantitative measurement methods (quantitative coronary arteriography): Condition following an anterior wall infarct and logged ventricle analysis (left); myocardial infarct (right).

Text functions

Text functions refer to the individual entry of various information:

- User-configurable image labeling
- Free annotation or text entries via defined text modules
- Comment lines for the image
- R/L display, etc.

9.3 Image Quality

Image quality is relatively difficult to measure since it depends on a large number of physical and physiological factors even when the geometric conditions are identical.

Image quality refers to the qualitative, diagnostically analyzable condition of the X-ray image. This will always be partially subjective. The radiation quality provides the foundation for good image quality and radiation hygiene.

- Image quality is defined in radiological terminology as the relationship between the structures of a test sample to be irradiated with X-rays and the parameters of its visualization.

A number of factors influence image quality both positively and negatively. These are:

- The object to be irradiated, i.e., its size, its thickness and density (different absorption for bones, tissue, organs, etc.).
- Movements (unsharpness or artifacts from own motion, system motion, electronic interference factors, subtraction artifacts, etc.).

9.3 Image Quality

- Exposure data, such as exposure voltage U_R (kV) and tube voltage I_R (mA) and exposure time t (s).
- Geometric exposure parameters, such as the focal spot size and distances.
- Scatter radiation reduction, such as collimation, filtering or compression of the object.
- Contrast agent.
- Physical properties of the imaging systems, such as the image intensifier, detector, film, screen (structure noise).
- Electronic microprocessor components (quantization noise).

9.3.1 Image quality requirements

The required image quality entails the reproduction of an artifact-free, high-resolution image of the region of interest with good contrast and low noise.

The image depicted in Figure 9.48 on the left shows four variables that negatively influence image quality, e.g., artifacts, poor contrast, high noise and low resolution, as negative features.

Figure 9.48-1 does not show any artifacts but continues to show poor contrast, high noise and low resolution. Figure 9.48-2 shows an additional improvement in the contrast, Figure 9.48-3 shows reduced noise, and Fig. 9.48-4 shows improved resolution. Figure 9.48-4 does not show any of these negative influences, i.e., it shows an acceptable exposure. Under the consideration of all imaging parameters, it meets the requirements for a high-quality diagnostic X-ray with organ-specific image features and details (the original high quality of the image cannot be properly shown in print).

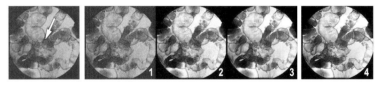

Figure 9.48 The image on the left shows an unacceptable exposure! It includes artifacts (arrow), poor contrast, high noise and low resolution. The right side shows individual improvement steps for obtaining a high-quality diagnostic exposure.

9.3.2 Image quality factors (characterizing parameters)

Despite the numerous influencing factors of varying importance, different parameters are available for defining image quality. These are:

- Artifacts
- Sharpness, image sharpness

9 X-ray Imaging Technology

- Contrast, image contrast
- Spatial resolution, detail resolution
- Modulation Transfer Function (MTF)
- Grayscale resolution, gradation
- Signal-to-noise ratio

Artifacts

Artifacts are visible interferences in the image and can have numerous causes. They are typically generated as "motion artifacts" by patient or system movements. They cannot always be prevented – not even by using good equipment and working carefully. Artifacts are not only generated in radiography. They can occur in all imaging systems.

Digital radiography provides operating modes or image postprocessing modes that, for example, prevent subtraction artifacts via a heart phase-controlled exposure mode or remove motion artifacts via pixel shift (see section 9.2).

Sharpness, image sharpness

Image sharpness parameters that negatively influence image sharpness are also referred to as image unsharpness or simply unsharpness. The following terms are used:

- Motion unsharpness
- Geometric unsharpness
- Focus unsharpness
- Film and screen unsharpness

Motion unsharpness

This is a result of patient, organ or imaging system movement. It is the unsharpness resulting from motion. This unsharpness may produce an artifact on the image.

However, there are also desired patient or system movements for blurring superposed details (resulting in unsharpness), e.g., for skull and cervical spine exposures, a.p. via the intentional movement of the lower jaw or during tomography (see section 9.4).

Geometric unsharpness

Geometric unsharpness is influenced by the distances between the focus and the imaging plane (see section 9.4). As the object's distance from the imaging plane decreases, the geometric unsharpness also decreases.

9.3 Image Quality

Focus unsharpness

The ideal focal spot (focus) would be punctiform. However, since a focal spot always encompasses a surface with certain dimensions (see sections 7.3 and 9.1), the different sizes of the focal spots result in different geometric unsharpnesses. As the focus increases, the geometric unsharpness also increases. The focus unsharpness is also a geometric unsharpness.

Figure 9.49 shows different geometric parameters, such as focus size (A), object size (B) and distances (A/B; B/C). They influence the degree of the geometric unsharpness (S1-S3).

The right side of Figure 9.49 shows details of objects (B) that are the same size or smaller than the focus (A). They are visualized smaller on the image receptor plane C1 with significant geometric unsharpness (S4), and on plane C2 only as unsharpness (S5). They are no longer visualized on plane C3. They are only reproduced as paradoxical shadows (S6).

Film and screen unsharpness

Film and screen unsharpness result from the sizes of the crystals, grains (also see structure noise) and their statistical distribution (graininess) as well as the scatter properties in the film-screen layers.

Light is emitted on all sides from the luminous crystals and is scattered to other crystals. Light that travels a longer distance within the thicker phosphor layer reaches the film with a light cone with a greater diameter. Crystals close to the surface have a shorter distance to travel to the X-ray film and naturally produce less screen unsharpness. A red or yellow binding agent absorbs more laterally obliquely scattered light. This slightly reduces the screen unsharpness however at the expense of the amplification.

Example: If an object with a diameter of 1mm is placed directly on a cassette (only with film, without an intensifying screen, i.e., screen unsharpness = 0), the object would also be visualized on the film with a diameter of 1 mm. Given a film-screen combination with screen unsharpness $U_{fol} = 0.2$ mm, this

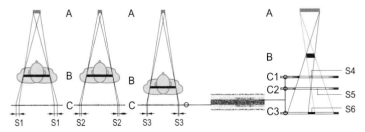

Figure 9.49 Schematic diagram of geometric unsharpness (left) and diagram of paradoxical shadow formation (right)

object would be visualized with an unsharpness edge of 0.2 mm, i.e., slightly magnified. This applies to all object details of an X-ray image.

Contrast, image contrast

To be able to diagnostically evaluate an image, one part must be brighter or darker than another image region. The visibility of the bright-dark difference between adjacent image locations (structures, pixels) is generally referred to as the contrast. This is a measure of the difference in luminous density between a detail and its direct surroundings.

A specific dose penetrating the matter forms a radiation contrast (also object area, object contrast range, object contrast) which is displayed as the density contrast on the X-ray image. The density contrasts of the X-ray image correspond with the radiation contrasts. The radiation contrast must be sufficient for imaging. It is defined as the ratio of the logarithm of the greatest to smallest dose in the imaging radiation cone.

Greater radiation contrast produces greater density differences in the image, i.e., it produces a greater radiographic range or exposure latitude and results in greater contrast (Fig. 9.50-B left). Images with low contrast (Fig. 9.50-B right) have fewer density differences (gray levels). This can be desirable for certain diagnostic situations.

- The maximum contrast of an image is the ratio of the brightest to the darkest location in the image. This is also referred to as the dynamic range.

There are a number of possibilities for the formulaic determination of density and luminous density differences. The numerical values of the same image can differ greatly depending on the formula used. Therefore, when providing the image contrast or the contrast transfer factor K, it is necessary to indicate the formula used. In general, K_2 should be specified.

If L_1 is the maximum value and L_2 is the minimum value of the density or luminous density, the following results:

- $K_2 = (L_1 - L_2)/(L_1 + L_2)$ = contrast transfer factor

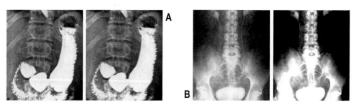

Figure 9.50 Example of different levels of image unsharpness in a stomach survey in a standing position (A) and an example with different image contrast in a pelvic radiograph (B)

Therefore, the contrast transfer factor K is always a value between 1 and 0. Other formulas for determining the contrast transfer factor are:

- $K_1 = L_1 / L_2$
- $K_3 = (L_1 - L_2) / L_1$
- $K_4 = (L_1 - L_2) / L_2$
- $K_5 = \log L_1 / L_2$
- $K_6 = L_1 - L_2$

Spatial resolution, detail resolution

Spatial resolution (detail resolution, local resolution, resolution, or resolving power) refers to the smallest discernible brightness difference between two adjacent pixels or object details. It is influenced by the entire image transfer chain (see modulation transfer function).

Next to contrast, sharpness and the signal-to-noise ratio, spatial resolution is one of the most important parameters for characterizing image quality. Line grids or lead line grids are used to measure (test) the resolving power of radiography systems using a direct radiography technique.

A line grid or lead line grid (Fig. 9.51-A) is used to measure deviations of object or image details. The lead line grid is positioned in front of the imaging system, e.g., on the patient table.

Three openings of identical width (bars, lines, strips) are made at equal distances in a thin lead sheet, e.g., 0.05 mm. They represent line pairs (one bright and one dark line). The values (0.8 to 5.4 in this example) indicate the number of line pairs per millimeter (LP/mm).

If lines and intermediate spaces can still be clearly identified in increasingly finer gradations (Fig. 9.51-B), the image (the image transmission system)

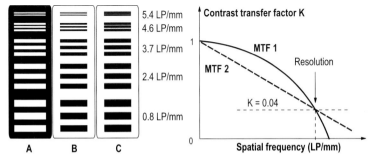

Figure 9.51 left: Basic diagram of a line grid for determining spatial resolution, spatial frequency. Right: Resolution (spatial resolution, spatial frequency) of two different image transmission systems.

has a high resolution, i.e., a high spatial frequency. If the lines and intermediate spaces are blurred, e.g., at 5.4 LP/mm (Fig. 9.51-C), the spatial frequency (resolution) is only 4.6 LP/mm.

The curve of the spatial frequency as a function of the transfer parameters is referred to as the modulation transfer function (MTF).

The right side of Figure 9.51 shows the spatial frequency (resolution) as a function of the contrast transfer factor K. The two image transmission systems have different MTFs. MTF1 is clearly better. However, the two MTFs intersect the (subjective) visibility threshold $K = 0.04$ at the same point and therefore have the same resolution. When measuring only the resolution, both systems would seem to be equal. However, this shows the importance of the highly diagnostic MTF. Test bodies for fluoroscopy systems have a similar design to those in the direct radiography technique. They also contain a lead strip test for testing the resolution, as well as five concentric, radiation-absorbing rings with different diameters of 120 mm to 288 mm.

Modulation transfer function (MTF)

The imaging properties of a system comprised of several components are calculated via the modulation transfer function (MTF). The X-ray image generated with optimal exposure parameters (requirement) is ideally to be visualized following penetration of the object without a loss of information as a diagnostically usable image.

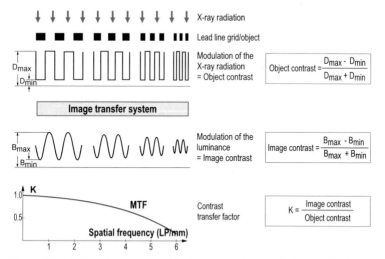

Figure 9.52 Schematic diagram and basic process of a modulation transfer function MTF. The measurable value of the modulation transfer function is an indication of the qualitative assessment of an imaging system (image transfer system, imaging system). It is determined from the amplitudes of the individual contrasts.

9.3 Image Quality

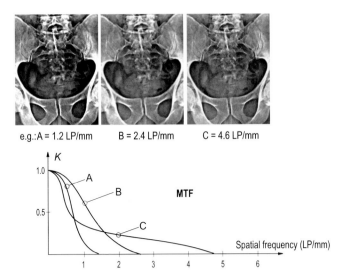

Figure 9.53 Basic diagrams which could be produced, for example, via imaging systems with different modulation transfer functions: Examples of the MTF and the image result. The measurable value of the MTF is an indication of the qualitative assessment of an imaging system (manipulated view).

Depending on the technology of the image transfer system, the X-ray image is converted into light, into electrons and/or electrical signals, or back into light. This means that in the process of becoming a visible image the X-ray image travels a relatively complex path through components for constantly converting and influencing the image information (Figs. 9.52 and 9.53).

The MTF is the representation of the contrast transfer factor K as a function of the spatial frequency of the object (test piece).

K is the ratio of the modulation of the optical density (for X-ray film) or the modulation of the luminous density (on the I.I. output screen or FD scintillator) to the modulation of the radiation at the input of the imaging system.

The relationship which describes the inertia (motion unsharpness) of imaging systems is referred to as the temporal modulation transfer function. Instead of the spatial frequency, a temporal series of dose rates changing in a sinusoidal manner is used in this process.

Grayscale resolution, gradation

Grayscale resolution refers to the maximum number of grayscale values that are displayed. In the digital exposure technique, it is a function of the bit depth (see section 9.2).

The gradation determines the relation of the gray levels to one another. Images are often changed by characteristic curves (look-up tables) during transmission. More significant differences in the input signal can be better transmitted and displayed, e.g., for an X-ray image of a thorax. For information on gradation, see section 8.1 and for look-up tables, see sections 8.1 and 9.2.

Signal-to-noise ratio

The signal-to-noise ratio, SNR, is the ratio of the effective values of the useful signal distribution to the statistically based interference signals (also see section 8.3).

Useful signals are the image-forming portions of the X-ray image. Interference signals are any interfering signals of the image-forming components and negatively effect image quality in the form of noise or image noise.

Image noise can be defined as a group of interference signals that are temporally or locally restricted or change constantly, are caused by statistical phenomena, and have no connection with the object to be visualized. Interference signals are sine portions (superposition) or larger frequency ranges.

A differentiation is made, for example, between:

- Quantum noise: Discontinuous impinging of X-ray quanta on the image transmission plane. In the case of a low dose rate, i.e., low amount of quanta per time unit, the quantum noise is particularly visible in the X-ray image since its statistical distribution is rougher. It reduces as the dose rate increases as a result of the increase in X-ray quanta in the useful beam.
- Structure noise (graininess): This occurs in photographic layers as a result of the statistical distribution of the grainy structure (noise as a grain pattern) or the statistical distribution of crystal structures.
- Quantization noise: Also called electronic noise. This occurs, for example, in the case of semiconductor intensifiers in TV cameras or analog-digital converters.

9.3.3 Measures for improving image quality

For digital imaging systems with image processing or postprocessing, there are possibilities for improving image noise, for example, in the form of image integration (see section 9.2). The dose or the contrast agent concentration may also be increased, for example.

The image noise is reduced by half when the dose is quadrupled or the contrast agent dose is doubled.

The necessary dose is regulated via the automatic exposure controls (section 9.1). However, some special conditions require a comparison of the exposure values obtained via functions such as transparency compensation (section

9.1). Object magnification via a change in geometric distances should be avoided as possible since they negatively affect image sharpness and detail resolution.

Scatter radiation is a significant factor in the negative influencing of image quality (section 3.2). To keep the scatter radiation as minimal as possible, there are different complementary solutions, e.g., grids and compression devices for reducing scatter radiation, filters for reducing low-energy X-rays, object collimation, etc., along the path of the X-rays from the focus to the image receptor medium.

Geometric magnification

Since there is always a geometric distance between the object to be recorded and the image receptor plane, the X-ray image is always slightly larger than the object to be visualized. The distance from the object to the image receptor plane depends on the additional components located in between, the possible exposure geometry, or the position desired for the exposure.

Regardless of the image receptor system, the ratio of the film-object distance (FOD) to the object-film distance (OFD) is to be as large as possible so that the magnification of the visualized object (E1) is negligible (Fig. 9.54). If the OFD increases or the FOD decreases, the visualized object (E2) is magnified with respect to the actual size (N).

The constructive design of radiography systems with integrated imaging systems has certain limitations. A standard source-image distance f_0 (SID) of 115 cm has proven to be effective in terms of the effect on the imaging geometry in radiography. Particularly in thorax exposures, i.e., exposures using bucky wall systems, the SID can be selected as desired. A SID of 150 or 180 cm has proven to be effective for this exposure technique.

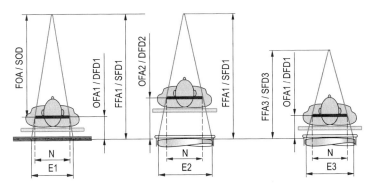

Figure 9.54 Magnification effect at different source-image distances (SID). I in SID stands for image and refers to the image receptor plane. This can also be the I.I. input screen, the imaging plate or the flat detector.

9 X-ray Imaging Technology

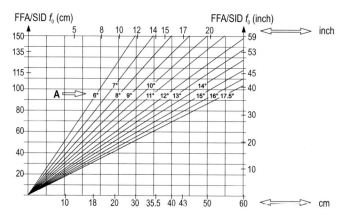

Figure 9.55 Illumination formats in cm/inch for different source-image distances (SID) in cm/in and anode inclination angles (A)

Undertable fluoroscopy systems or C-arm systems for angiography and surgery necessarily generate a magnified representation of the image due to their geometric ratios of FOD to OFD. Undertable fluoroscopy systems have an SID between 65 and 100 cm. In the case of C-arms for angiography, the SID can be set between approximately 90 and 120 cm, and for C-arms for surgery, the SID is approximately in the range of 90 cm.

Particularly in undertable fluoroscopy systems, the SID can be relatively small. Object magnification due to a larger angle of the radiation cone is unavoidable. A larger angle of the radiation cone is necessary to be able to visualize an object, e.g., up to 35 cm × 35 cm, with the defined geometric proportions. X-ray tubes with a greater anode angle are used to achieve this (see chapter 7).

Figure 9.55 shows the relationship of the source-image distances f_0 (SID) and the maximum geometric illumination for generating an image as a function of the anode inclination angle.

Ideally the object would be visualized with a 1:1 scale. This is not possible due to the geometric proportions (see Fig. 9.54 and section 9.4). To create a visualization with as little magnification as possible,

- The object should always be positioned close to the image receptor plane.
- The source-image distance should be selected to be as large as possible.

Magnification via distances ratios should be avoided since they always negatively effect the image sharpness and detail resolution (see chapter 9.3). For improved detail visualization, organs must nonetheless be able to be displayed with magnification.

9.3 Image Quality

There are different algorithms for this that magnify an object or an image segment in analog as well as digital image generation or image reproduction systems.

In the case of image intensifiers, the switch to a smaller image input format on the output fluorescent screen yields a magnified visualization of the recorded object. This magnification is produced by switching the formats, i.e., using a smaller diameter of the real I.I. input format. Only the smaller object to be visualized is actually transmitted to the same area of the I.I. output screen. Reference is made in this context to format changeover, zoom formats or zoom levels for electron-optical magnification (see section 8.3).

Given unfavorable distances, the selection of a smaller focal spot f (focus) can reduce at least the focus unsharpness (example Fig. 9.56). However, a smaller focus does not have as high a loading capacity.

The magnification M of an image and the focus size f influence the limiting resolution N. This can be approximately shown as a function of the changeable geometric exposure relationships using the two diagrams (Fig. 9.56).

Example for Fig. 9.56: Given a source-image distance f_0 (SID) of 85 cm and an object-film distance (OFD) of 25 cm, the magnification is $M = 1.4$ (top

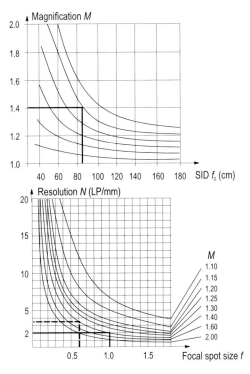

Figure 9.56
Relationship between geometric magnification and spatial resolution

diagram). Given a magnification of $M = 1.4$ and the use of a focus size f of 1.0, the resolution is $N = 2$ LP/mm (bottom diagram). When switching to the focus size of 0.6, the resolution N improves to 3.5 LP/mm.

Scatter radiation grid

When penetrating the object, X-ray quanta interact with the atoms of the object-specific matter (see chapter 1). The resulting scatter radiation (primarily Compton radiation) has a significant negative effect on image quality, i.e., the contrast. As the absorbed scatter radiation increases, the positive effect of the direct radiation on image quality increases. The scatter radiation grid is a component for absorbing scatter radiation.

The image-forming primary radiation coming from the focus of the X-ray tube in a straight line travels unattenuated behind the object through the interspace material to the image receptor system. The scatter radiation emitted by the object (and other matter), also referred to as secondary radiation, is largely absorbed by the grid (Fig. 9.57).

The honeycomb grid used for the first time in 1912 by Gustav Bucky to reduce scatter radiation (also see chapter 7.5) continues to be of value. The honeycomb grid became a highly sensitive device, the grid: Latin raster, rastrum.

Figure 9.58 shows the structure of a grid for reducing scatter radiation: Scatter radiation grids are made of a series of very thin parallel lead plates, known as absorber leaves (R1), that are tilted toward the focus either edgewise (R2) or symmetrically. Lead as a chemical element with its atomic number of 82 absorbs almost all radiation (quanta) relevant for radiodiagnostics. Large quantities of radiolucent materials, such as paper or aluminum, are used as the intermediate layer, known as the interspace material. The

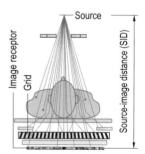

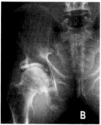

Figure 9.57 Scatter radiation reduction using a grid: The exposure A of a pelvic phantom was made without a grid with 75 kV. It shows substantial scatter radiation (significant fogging). The exposure B is of the same object with 75 kV but using a scatter radiation grid with grid ratio of 12.

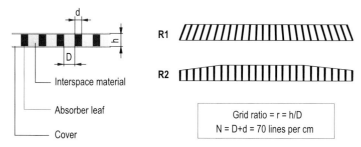

Figure 9.58 Grid structure, dimensions and grid type

absorber leaves and interspace material have a top and bottom cover to form a connection and to protect them from damage.

R1 and R2 show sections of a focused grid (R1) and a parallel grid (R2). In the case of focused grids, all lead leaves are to point toward the focus. In the case of large exposure formats, the image-forming radiation is absorbed more strongly at the edge in parallel grids. To prevent this, parallel grids are slightly tilted toward the edge.

Definitions and data

The geometric parameters of a grid are:

- D = distance of the absorber leaves – corresponds to the thickness (width) of the interspace material.
- d = thickness of the absorber leaves.
- h = height of the absorber leaves and the interspace material, approximately 1.5 mm.
- $N = D + d$, number of lines per cm, e.g., 70 lines per cm.
- $r = h/D$, grid ratio, e.g., 17/1, written as: 17:1.
- f_0 = focusing range, e.g., for 150 cm SID.

Data and definitions are provided in grid data sheets: e.g., Pb 17/70; f_0 150, indicates the following:

- 17 indicates that the grid ratio is $h/D = 17:1$. 70 indicates that the number of lines is $D + d = 70$ per cm.
- Pb is the chemical symbol for lead and means that the absorber leaves are made of lead.
- f_0 150 indicates the ideal source-image distance (SID), 150 cm in this instance. For information on focusing limits, see the section on grid focusing and Figure 9.59.

The following example shows how highly sensitive and precise a grid is: Grid Pb 17/70; f_0 150 signifies that for this grid the width d of an absorber leaf is 0.11345 mm and the width D of an interspace material is 0.02941 mm.

Multi-line grid

Stationary, i.e., fixed, grids with a smaller grid ratio and a relatively low line number would visualize grid lines on the image. As a result, such grids are moved during exposure in the transverse direction with respect to the lines (typically approximately 1 cm).

Grids with a high grid ratio (e.g., 17:1) and a relatively high number of lines do not need to be moved during exposure since they do not visualize any interfering or noticeable grid lines. Starting at 60 lines per cm, they are referred to as multi-line grids.

Grid cassettes

Grid cassettes are employed for the use of scatter radiation grids in application cases in which a radiography or imaging system with an installed grid is not available. Grid cassettes have a fixedly installed parallel grid, i.e., with parallel lines.

Circular grid

Circular grids are positioned in front of the input screen of X-ray image intensifiers. In contrast to quadratic scatter radiation grids, these are circular. Their inner structure corresponds fully to the quadratic scatter radiation grid with parallel lines.

Criss-cross grid

The less common criss-cross grids are comprised of two parallel grids which are twisted against each other by 90° and are connected to form a single grid. Criss-cross grids are not suitable for exposures using oblique irradiation. They are very sensitive to any decentering.

Grid position

Grids are positioned so that the lines run in the patient's longitudinal direction so that the grid lines are blurred even in conventional linear tomography, for example (see section 9.4).

The transverse movement of the grid by approximately 1 cm with respect to the patient's longitudinal axis is started when the exposure is triggered. The grid movement can be either a one time occurrence (moving grid) or an oscillating occurrence (oscillating grid). The term "catapult" which is known in connection with the term catapult bucky tray comes from the catapult-like

9.3 Image Quality

start of the grid movement. As a result, a bucky tray with a stationary grid cannot be a catapult bucky tray.

In the case of X-ray systems with only image intensifiers or flat detectors as the imaging system, e.g., all C-arm systems in angiography or surgery, multi-line, stationary and/or exchangeable grids are positioned in front of the imaging medium.

Exposure extension via grids

Grids mainly absorb scatter radiation but also absorb direct radiation. The grid structure clearly illustrates this. The percentage of scatter radiation in the case of thick objects and large exposure fields can be greater that 80%, i.e., in such a case the image-forming direct radiation is less than 20%.

Example:

If during an exposure without a grid the dose on the image receptor medium is assumed to be 100, the value when using a grid is only approximately 16 depending on the grid structure. This means that although the exposure using a grid requires a 6.25x greater dose (referred to as the exposure extension factor), the composition of the image-forming radiation is significantly better. The scatter radiation percentage is only approximately 25%.

Exposure without a grid:
Total radiation   100%
Scatter radiation   > 80%
Primary radiation   < 20%

Exposure with a grid:
Total radiation   approx. 16%   corresponds to 100%
Scatter radiation   approx. 4%   corresponds to 25%
Primary radiation   approx. 12%   corresponds to 75%

Selectivity S

Depending on the grid type, the percentage of image-forming direct radiation can be improved. The selectivity S characterizes the effectiveness of a grid. It is defined as the ratio of the primary radiation permeability to the secondary or scatter radiation permeability.

Example:

If 30% of the direct radiation is absorbed by the grid, the image-forming direct radiation is 70%. The primary radiation permeability is thus 0.7. If 90% of the scatter radiation is absorbed by the grid, the image-forming scatter radiation is only 10%. The scatter radiation permeability is thus 0.1.

This yields a selectivity S: Primary radiation permeability/scatter radiation permeability: $S = 0.7/0.1 = 7$. A scatter radiation permeability of 0 would be

9 X-ray Imaging Technology

ideal but is not possible since the selectivity value would have to equal infinity.

Harder scatter radiation is absorbed less by thin lead leaves than soft (see section 3.4 – hard and soft X-rays). With higher tube voltages U_R, i.e., with higher kV values, the selectivity value increases. The gray levels in the image increase correspondingly.

Lead content of grids

The lead content p (the lead amount) of a grid can be of interest for fundamental considerations but is not important for practical applications. The lead content relates to a unit length l_0 of 1cm or 1mm and is defined as

$$p = d \cdot h \cdot N/l_0.$$

Example: For a grid $d = 0.07$ mm (thickness of the lead leaves), $D = 0.18$ mm (thickness of the interspace material) and $h = 1.4$ mm (height of the lead leaves), the line number is calculated to be $N = 4$ lines/mm. Therefore, the following results for the lead content $p = 0.07$ mm $\cdot 1.4$ mm $\cdot 4$ mm$^{-1} \cdot 1$ mm^{-1} = approx. 0.4; i.e., 40%.

Grids can have the same lead content regardless of the grid ratio and the line numbers.

Grid focusing

In connection with scatter radiation grids, grid focusing (also focusing) means the alignment of the lead leaves toward the focus. The focus must always be vertically centered on the grid center. Incorrect focusing (defocusing) results in the loss of image-forming primary radiation.

The effectiveness of a focused grid (Fig. 9.59) is significantly limited when

- It is decentered, i.e., the focus is not in the plane of the center leaf (B),
- It is defocused, i.e., a SID is set outside of the tolerances (C).

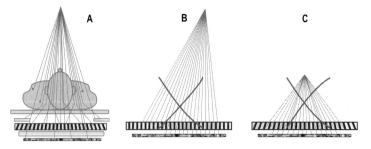

Figure 9.59 A schematically shows correct focusing and the ideal source-image distance (SID). B and C show incorrect grid focusing.

9.3 Image Quality

The primary radiation is absorbed significantly already at minimal geometric deviations. As a result, the dose requirement increases, the selectivity decreases, and the exposure has less contrast.

f_0 150 cm as the ideal source-image distance (SID) can be easily changed within certain thresholds. Thresholds refer to the smallest and to the greatest SID at which the absorber leaves allow the primary radiation pass through the grid opening (the interspace material) undisturbed. In the example "grid 17/70 f_0 150" this means that the grid is focused at a SID of 128 to 182 cm.

The thresholds of the usable source-image distances depend on the grid ratio and the absorber leaf number (line pairs per cm). Given the same source-image distance of f_0 115, for example, the thresholds differ as a function of the grid ratio and line pairs/cm:

f_0 115, 8/40: from 89 cm to 161 cm

f_0 115, 12/40: from 97 cm to 142 cm

f_0 115, 17/70: from 101 cm to 133 cm

f_0 115, 15/80: from 100 cm to 136 cm

Compression

An important aspect of scatter radiation reduction is the compression of tissue structures. Compression changes object density, and less scatter radiation is produced (Fig. 9.60).

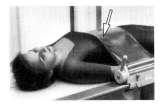

Figure 9.60
Compression via a belt compressor for reducing the object density and consequently the scatter radiation

Other possibilities for object compression include, for example, flat tubes for the over and undertable fluoroscopy systems which are applied to the appropriate object location. Mammography exposures are almost impossible without compression since the high percentage of scatter radiation in the female breast would not yield an effective diagnostic image.

Collimation

With respect to Figure 9.61: After the radiation cone is emitted from the tube, the radiation field in the primary collimator is collimated to the diagnostically relevant surface, also collimation near the focus; exclusively the iris diaphragm in the case of the I.I. (Fig. 9.61-C1-D).

9 X-ray Imaging Technology

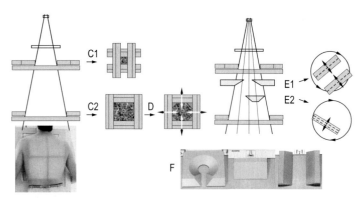

Figure 9.61 Collimators and preliminary filters near the focus and the film

Preliminary filters provide object-specific filtering, e.g., for bone-tissue transitions (E1 and E2). Compensating filters can be inserted as needed in the accessory rails of the multileaf collimator for density compensation during exposures in the pelvic, foot, shoulder, thoracic part of the spine, lumbar vertebral column or skull region (F).

Radiographic and universal fluoroscopy systems have the possibility of realizing collimation near the film in their spotfilm devices or bucky trays. Collimation near the film protects against reflecting scatter radiation in the edge zones of exposures, particularly in the case of format division.

Digital procedures

Special operating modes of digital radiography, such as filtering, DSA, etc., as well as a number of digital image processing algorithms significantly improve image quality as described in section 9.2.

9.3.4 Contrast agent

X-ray images are summation images (see section 9.4 – superposition) and due to the low density differences in body tissue (soft tissues) individual organs or organ parts cannot be differentiated or only with difficulty. For information on contrast agent examinations, see chapter 6.

To be able to make a sound diagnosis, i.e., to achieve high image quality, the density differences must be made visible in the radiographic examination. This is achieved by introducing certain agents (chemical elements) to create an "artificial" contrast (Fig. 9.62). As a rule, two types of substances can be used for such agents, known as contrast agents:

9.3 Image Quality

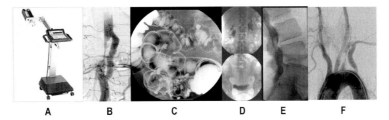

A   B   C   D   E   F

Figure 9.62 X-ray imaging using contrast agents. A: Contrast agent injector. B: Vessel segment with bypass. C: Double contrast colon exposure. D: Excretory urogram. E: Myelography, spinal marrow exposure. F: Aortic arch.

- Contrast agents of barium sulfate compounds, such as $BaSO_4$ (atomic number of barium Ba = 56, of sulfur S = 16) and primarily organic iodine compounds (atomic number of iodine = 53) since they attenuate or absorb the X-rays better than the surroundings. They are also referred to as "X-ray positive" contrast agents.
- Contrast agents of gases or air, such as CO_2, which attenuate or absorb the X-rays less than the surroundings (atomic number of carbon C = 6, oxygen O = 8). They are also referred to as "X-ray negative" contrast agents.

Since soluble barium compounds are extremely poisonous to human organisms, only the insoluble sulfuric acid salt of barium is used as the contrast agent. Iodine compounds can induce incompatibility reactions in some cases in arterial or intravenous injections.

There are different manners for introducing the contrast agent into the individual organs:

- Via puncture, e.g., for visualizing vessels or cerebral ventricles.
- Via natural access, e.g., mouth, rectum, urethra.
- By directing the contrast agent via an organ function (indirectly), e.g., the function of the liver to visualize the gallbladder and common bile ducts or the function of the kidneys to visualize the excretory urinary tracts.

The contrast agent is typically expelled naturally (gastro-intestinal tract, ureter, lungs).

The sectional image technique, computed tomography (CT), is also used for the radiological imaging of organs and vessels in the skull, thorax and abdomen region.

Double contrast

Gastro-intestinal examinations or joint exposures are often performed using a combination of both contrast agent types. After the colon, for example, is filled with a barium sulfate paste via a barium enema and is then emptied again, paste residue remains attached to the colon wall. If the colon is then filled with air, the air together with the paste on the colon wall yield a double contrast (Fig. 9.62-C).

Contrast agent injectors

Contrast agent injectors are used in angiography to control contrast agent administration (Fig. 9.62-A). For direct operation of all functions in the examination room, injectors can be positioned on a movable unit at any location of the patient table or directly at the patient table. They are used to control the contrast agent pressure or flow rate.

9.4 Projections and Imaging Planes

An X-ray image is created by projecting body parts onto an image receptor. The imaging planes necessary for diagnosis result from the projection.

9.4.1 Projections

All objects located in the radiation cone are projected onto the image receptor plane (superposition). The blackening of the exposure results from the superposition of all objects located in the imaging plane as a function of absorption and scatter.

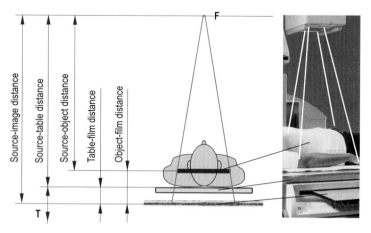

Figure 9.63 Terms for the various distances defined for the exposure geometry

The word film represents the image receptor plane. This can also be the I.I. input screen, the imaging plate or the flat detector. The standard projection illustrated in Figure 9.63 is used to describe the distances that affect the geometric visualization of the image. In addition to the distances between the focus, object and image receptor plane, these include the size of the focal spot (focus), the size of the object (collimation of the radiation cone) and the beam projection angle (perpendicular or diagonal to the object).

The work height of the examiner, i.e., the distance from the floor to the patient table, is not important for the exposure geometry but is important for ergonomic patient work (Fig. 9.63-T).

Superposition, oblique projections

Superposition is the overlaying of structures with different object depths. Since there is object superposition in every projection, exposures are made from two or three different angle settings or with an oblique or lateral projection to ensure effective diagnostic information (Fig. 9.64).

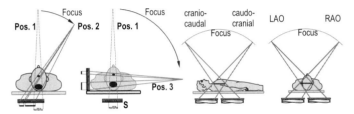

Figure 9.64 Frontal, sagittal (pos. 1), oblique (pos. 2) and lateral (pos. 3) beam path and oblique irradiation in the patient's longitudinal direction, orbital about the patient's longitudinal axis

Therefore in order to eliminate undesired superposition of objects (Fig. 9.64-S) and obtain diagnostically relevant findings, it is necessary to set the projection of the central ray according to the anatomic conditions.

Oblique irradiation (cranio-caudal and caudo-cranial) can be achieved with almost every X-ray system. C-arm systems, as used in angiography or in the OR, allow the above-mentioned projections as well as orbital angle settings (LAO and RAO) of 0 to 180° around the patient (orbital for loops, e.g., also around the atomic nucleus).

Double oblique projections

Double oblique projections are used in particular in cardiac angiography and neural angiography. They combine the two oblique projection possibilities (Fig. 9.65).

9 X-ray Imaging Technology

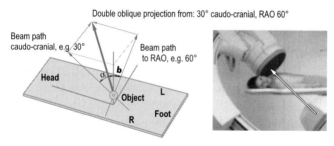

Figure 9.65 Schematic diagram of a double oblique projection with beam path from bottom to top

In this context, the oblique projections are on the transverse planes, i.e., the beam path runs axially, vertically or obliquely to the patient's center axis.

Cranial	belonging to the head, in the crest or head direction (from krani, cranio meaning skull)
Caudal	at the foot or tail (from kauda, cauda, tail)
Caudo-cranial	from the foot to the head
Cranio-caudal	from the head to the feet
RAO	right anterior oblique (also fencer position)
LAO	left anterior oblique (also boxer position)
anterior	Lat., front
oblique	from Latin, obliquitas, inclined direction

Conventional slice images

A special form of oblique projection is provided by conventional tomography (-tom, Greek from tome, slice; also tomo-, meaning cutting or segment).

The disadvantages of superposition are avoided in a slice image, tomography. An almost superposition-free image of all object details within a certain slice (thickness) is generated (Fig. 9.66). However, superposition can occur within the generated imaging slice in conventional linear tomography. This can be eliminated in CT or rotational exposures.

X-ray tube (focus F) and image receptor (Fig. 9.66-B) are mechanically coupled or coupled via electronic controls so that they shift in opposite directions during an angle range sequence (α = tomo angle). The position of the center of rotation (Fig. 9.66-A) corresponds to the position of the body slice to be visualized (a = tomo height).

9.4 Projections and Imaging Planes

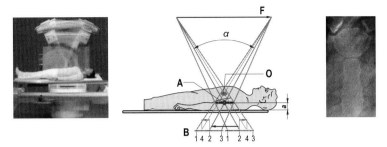

Figure 9.66 Linear tomography (planigraphy here). Tomo sequence on the left and result image on the right.

Objects (e.g., Fig. 9.66-O) over or under the set tomo height are blurred on the image receptor. The image shows that the sharply visualized image parts 1-3 hit exactly the same point of the image receptor during the entire movement while the superposed object (Fig. 9.66-4) does not and is blurred.

The thickness of the almost superposition-free slice depends on the tomo angle α. Imaging planes with slice thicknesses of approximately 2 to 10 mm can be created. As the tomo angle α decreases, the visualized body slice becomes thicker. In the case of tomo angles of approximately 8° to 10°, the term zonography is used. In standardized linear tomography, the tomo angle is approximately 40°.

A principle of tomography is also used in CT or rotational imaging. The imaging system rotates in this process orbitally, continuously, or also in an angle-controlled manner about the patient's longitudinal axis. Digital algorithms are used to generate images of superposition-free body slices from the transverse plane of the patient.

A continuous or angle-controlled movement of the imaging system is a requirement for rotational angiography or for image reconstruction for the three-dimensional imaging of objects. In CT and for X-ray systems with C-arms, this is achieved in angiography and surgery.

Today, only linear slice sequences are used in conventional tomography. At the highpoint of conventional tomography (over 30 years ago), other geometric sequences (multi-dimensional blurring) were practiced for visualizing superposition-free body slices, e.g., circular, elliptical, spiral, hypocycloidal. Hypocycloidal refers to tricyclic blurring in an egg shape.

Panoramic slice images

Panoramic tomography is often used in dentistry to visualize the teeth and middle third of the face region.

9 X-ray Imaging Technology

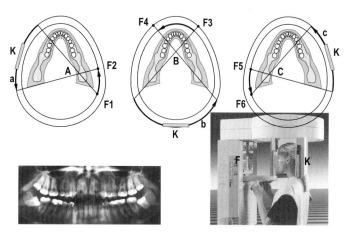

Figure 9.67
The tomo sequence in a panoramic slice images, imaging system and result image

Panoramic tomography is based on the principle of the previously described conventional tomography. Figure 9.67 shows the principle procedure: The X-ray tube is initially in the right position (F1). When radiation is switched on, the focus runs around the jaw region to the left position (F6). Coupled with the X-ray tube, the image recorder (K) simultaneously moves from the left to the right position behind the skull. The position of the tomo pivot point is changed on an ongoing basis so that it is almost always in the row of teeth.

The gray jaw region is to be visualized: The focus travels a path from F1 to F2, from F3 to F4, and from F5 to F6. The opposite image receptor (K) also moves in this process and travel the paths a, b and c. The tomo pivot points switch from A to B and to C. The shift of the point of rotation during the tomo sequence results in visualization of the individual segments of the tooth arch. Points of rotation A and C form the less curved slices of the side regions and B forms the more curved front tooth region.

Isocenter

In the case of C-arm systems as used in surgical radiography and angiography, the imaging system is arranged so that it is geometrically tangential to a circle (circle segment). The central ray always travels through the isocenter of this geometric circle regardless of the orbital angle. This center of rotation is referred to as the isocenter (Fig. 9.68, right). The advantage of isocentric object positioning is that the object stays in the same place on the monitor and the height position of the C-arm system does not have to be adjusted for all angular movements (projections) and not just orbital ones.

9.4 Projections and Imaging Planes

Figure 9.68 Projection of the two stereotactic imaging planes and result images: Stereo exposures (top) and needle marking (bottom). Right: Representation of the isocenter (iso).

Stereotactic biopsy

In mammography the exact location of the region of interest must be determined prior to a biopsy. This can be precisely calculated via stereotactic biopsy. Two exposures are made from two angle position, e.g., 16° in relation to the vertical beam path (Fig. 9.68).

The computer of the stereotactic biopsy accessory precisely calculates the spatial coordinates of the region of interest from the two exposures.

The precision mechanics for setting the needle holder is coupled with electronics so that the inserted needle point is exactly in the location to be biopsied.

Other exposure techniques

The following exposure techniques are no longer used today. The most frequently used method was the simultaneous multi-section method. The cross-sectional scan and stereo technology have not become established in radiology. However, for the sake of completeness, the three exposure procedures are addressed briefly.

Simultaneous multi-section method

In the simultaneous multi-section method, the sequence for creating the slice images is as in the already described conventional tomography. Instead of just one image receptor with a film-screen system, up to eight film-screen combinations were placed over one another in a cassette. These were simultaneously exposed so that exposures were created of different superposed organ slices with only one tomo sequence.

Cross-sectional scan

In the cross-sectional scan, transverse body slices typically in the thorax region were visualized. The central radiation beam was aligned in a stationary manner at an angle of approximately 25° from the horizontal in a downward direction toward a horizontally positioned cassette. During exposure, the sitting patient and the cassette were rotated 360° in the same direction about a vertical axis.

Stereo technology

Comparable to the technology of the stereotactic biopsy for mammography, a spatial impression of the exposure was made during stereoradiography. Two exposures were made of an object in quick succession with a focus shift of approximately 7 cm. Stereo glasses or stereo viewing devices were used for viewing (diagnosis). In this process, every eye saw only "its" image and generated in this manner a visual spatial image.

9.4.2 Imaging planes

Figure 9.69 shows different imaging planes (A, B, C and D), which visualized the object perpendicular to the central ray.

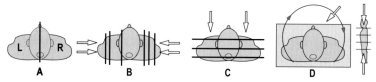

Figure 9.69 Imaging planes

A   The median plane (medial, toward the center of the patient) is the sagittal plane (sagittal from sagitta, arrow) running through the center of the body. The median plane (median sagittal) divides the body in the ventral dorsal direction into two equal parts; left (L) and right (R).

B   The sagittal planes are all parallel planes of the middle plane (median plane, sagittal plane) of the body or the sagittal suture of the skull. A lateral beam path produces exposures on the sagittal planes.

C   The frontal planes are all planes of the body perpendicular to the sagittal planes. The term frontal beam path ("frons", forehead) is used since the projections are parallel to the forehead, i.e., parallel to the patient table when the patient is prone or recumbent. A frontal (sagittal) beam path produces exposures on the frontal planes (a.p./p.a.).

D   Transverse planes are all planes perpendicular to the body's longitudinal axis. Projections with an axial beam path produce exposures on the transverse body planes. The images on the transverse planes are typical of those of the CT and rotational exposure techniques.

9.4 Projections and Imaging Planes

Sagittal beam path

Projections with a sagittal beam path produce exposures on the frontal body planes (Fig. 9.70).

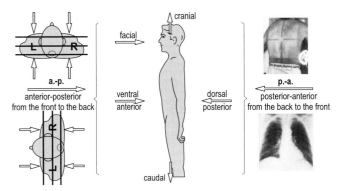

Figure 9.70 Sagittal beam path

anterior	from the front (ante- meaning before, anterior, front).
posterior	from the back (post- meaning after, back, later; posterior, the back).
a.p.	anterior-posterior, from front to back.
p.a.	posterior-anterior, from back to front.
ventral	abdominal-related (from ventralis, abdominal-related, belonging to the abdomen).
dorsal	back-related (from dorsum, backside, back; dorsalis, lying on the back).
v.d.	ventrodorsal, from the abdomen to the back.
d.v.	dorsoventral, from the back to the abdomen.
facial	face-related (belonging to the face, from fazialis, i.e., Nervus fazialis, the facial nerve).
cranial	belonging to the head, in the crest or head direction (from krani, cranio meaning skull).
caudal	at the foot or tail (from kauda, cauda, tail).

Side information left (L) or right (R) is always based on the patient.

9 X-ray Imaging Technology

Lateral beam path

Projections with a lateral beam path produce exposures on the sagittal body planes (Fig. 9.71).

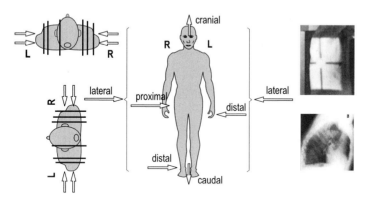

Figure 9.71 Lateral beam path

lateral	side (from lateralis, on the side).
proximal	near the body (from proximus, very close; parts of the extremities near the trunk).
distal	away from the body (from distalis, parts of the extremities further away from the trunk).
cranial	belonging to the head, in the crest or head direction (from krani, cranio meaning skull).
caudal	at the foot or tail (from kauda, cauda, tail).

Side information left (L) or right (R) is always based on the patient.

Axial beam path

Projections with an axial beam path produce exposures on the transverse body planes (Fig. 9.72).

axial (ax.)	in the direction of the body axis or the joints (from axis; also the second cervical vertebra).
volar	belonging to the palm of the hand (vola, hand surface; volaris, lying on the palm).
plantar	belonging to the sole of the foot (anat. planta, sole of the foot).

9.4 Projections and Imaging Planes

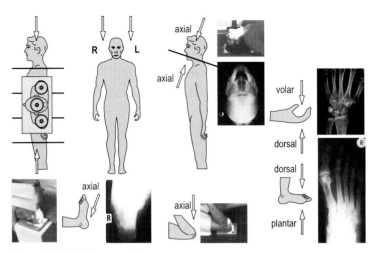

Figure 9.72 Axial beam path

dorsal	back-related (dorsum, backside, back; dorsalis, lying on the back).
volo-dorsal	from the palm to the back of the hand.
dorso-volar	from the back of the hand to the palm.
dorso-plantar	from the back of the foot to the sole of the foot.
planto-dorsal	from the sole of the foot to the back of the foot.

Side information left (L) or right (R) is always based on the patient.

Oblique diameters

Oblique diameters result from projection directions at which the body is obliquely irradiated on a transverse plane (Fig. 9.73).

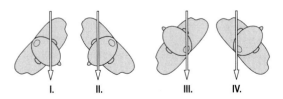

Figure 9.73 Oblique diameters. Oblique positioning of the patient or oblique projections are defined by the four oblique diameters: I. Oblique diameter, also "fencer position"; II. Oblique diameter, also "boxer position"; III. Oblique diameter; IV. Oblique diameter.

273

9.4.3 Radiological positioning method

The quality of the X-ray image is determined by the visualization of the diagnostically important image features, details, and structures that can be achieved with a medically tolerable low dose.

To achieve this goal, there are quality criteria that entail radiographic guidelines and physical parameters of the imaging system. Exposure guides, imaging references, setting manuals, etc., comprehensively describe and illustrate all radiographic projections under consideration of these quality criteria as well as the physical and geometric parameters (Fig. 9.74).

Literature on the radiological positioning method is available from the manufacturers and is also available in the technical literature; see references in the appendix.

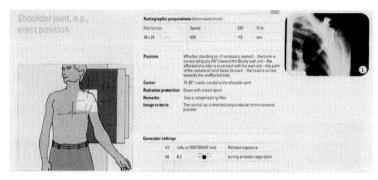

Figure 9.74 Example of the production of an X-ray image using an exposure guide

9.4 Projections and Imaging Planes

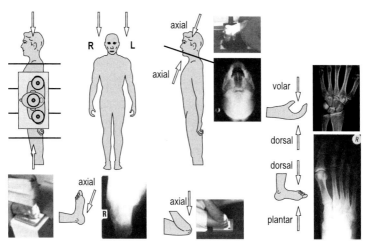

Figure 9.72 Axial beam path

dorsal   back-related (dorsum, backside, back; dorsalis, lying on the back).

volo-dorsal   from the palm to the back of the hand.

dorso-volar   from the back of the hand to the palm.

dorso-plantar   from the back of the foot to the sole of the foot.

planto-dorsal   from the sole of the foot to the back of the foot.

Side information left (L) or right (R) is always based on the patient.

Oblique diameters

Oblique diameters result from projection directions at which the body is obliquely irradiated on a transverse plane (Fig. 9.73).

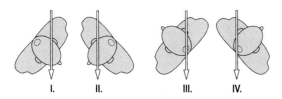

Figure 9.73 Oblique diameters. Oblique positioning of the patient or oblique projections are defined by the four oblique diameters: I. Oblique diameter, also "fencer position"; II. Oblique diameter, also "boxer position"; III. Oblique diameter; IV. Oblique diameter.

273

9 X-ray Imaging Technology

9.4.3 Radiological positioning method

The quality of the X-ray image is determined by the visualization of the diagnostically important image features, details, and structures that can be achieved with a medically tolerable low dose.

To achieve this goal, there are quality criteria that entail radiographic guidelines and physical parameters of the imaging system. Exposure guides, imaging references, setting manuals, etc., comprehensively describe and illustrate all radiographic projections under consideration of these quality criteria as well as the physical and geometric parameters (Fig. 9.74).

Literature on the radiological positioning method is available from the manufacturers and is also available in the technical literature; see references in the appendix.

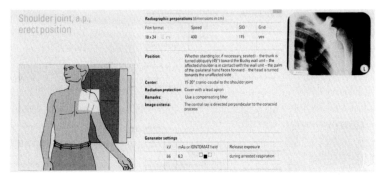

Figure 9.74 Example of the production of an X-ray image using an exposure guide

10 Patient Data Management

Medical recordkeeping is as old as medicine itself. Patient information, anamnesis (past medical history), and so forth were always documented and archived in some form, and with the emergence of radiology, radiographs and image data began to be collected.

10.1 Saving and Archiving Patient Data

Hand-marked X-ray film and other manual notes were the precursors to the modern patient record. The oldest method to record patient data on film (still in use today in one form or another) involves using a strip of paper and a simple device in the darkroom to expose the data on the edge of the film. Today, patient data is recorded on film in daylight using window cassettes and exposure cameras.

As in the past, data such as image projections, e.g., left or right, p.a./a.p., etc., are recorded on the X-ray using lead letters placed on the cassette or swung into the imaging beam field by hand or automatically. If needed, hand-written notes can be added to the X-ray, known as free annotation in digital imaging technology.

Electronic archiving of digital images is not yet explicitly regulated by law. Digital images can be copied to film using hardcopy or laser cameras (Fig. 10.1).

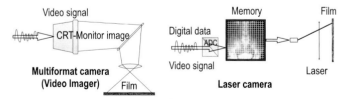

Figure 10.1 Schematic diagrams. Left: Function of a multiformat camera (video imager). The analog video image signal is reproduced and photographed. Right: Function of a laser camera (laser imager). Digital image data are copied directly into the memory of a laser camera; analog data are copied using an AD converter. A laser beam copies the data to film (right).

10 Patient Data Management

- Permanent film copies of monitor images can be created using a multiformat camera.
- A laser camera (laser imager) can be used to reproduce monitor images and/or digitally stored images on film, or images can be copied to paper and then printed (laser printer).

X-ray images can also be converted into digital signals and stored electronically using a film digitizer for digital processing and archiving. The film and its spatial gray-scale distribution is scanned one line at a time using an electron beam. The resulting analog voltage signals are digitized and are then available electronically for documentation, post-processing and archiving.

Once digital imaging systems were introduced, patient data were entered using a keyboard. A video character generator converted letters and numbers into electrical signals, which were included in the electronic image information. Electronic recording of patient data remains unchanged in principle. Regardless of the technology of the image acquisition system, both film and digital images must be documented and archived with standard and variable data. Archiving regulations are country-specific. In Germany, the data must be kept for 10 years. The statute of limitations actually expires after 30 years. According to copyright law, the radiograph is and remains the property of the imaging hospital/physician.

10.2 Patient File

All patient-related data must be documented and stored in a patient record in the form of a ring binder or file folder. The patient record may fill several folders, depending on past medical history (anamnesis), the frequency of exams, treatment, etc.

Finding space for and creating easy access to these conventional medical records is problematic enough for the practicing physician, let alone for a large hospital.

As digitization advanced, however, data began to be recorded from a variety of sources and then shared electronically. As a result, a kind of "hybrid recordkeeping" exists today, a combination or mixture of conventional and electronic patient records.

Even though conventional data collection and archiving methods are still in use, digital imaging spawned tremendous change in patient data management through medical information science, which continues to change and grow.

The contents (notes) of patient records, as anonymous electronically stored data, significantly help to optimize medical reporting and better the healthcare system. In addition, efficient communication would be unthinkable today without the vast electronic store of medical knowledge, e.g., in the form of databases, technical literature, etc. In spite of the controversy that surrounds the sensitivity of medical records, they do benefit the individual and society.

10.3 Medical Information Science

Computer science is the science of electronic data processing systems and the basic principles of their application. Since the 1960s, it has evolved to become a new basic science.

Medical information science supports the systematically applied data processing, documentation, and archiving of comprehensive medical data. It has been considered a separate clinical discipline since the 1970s.

The complexity of medical/ecological relationships continues to place great demands and growing responsibility on all users and manufacturers working in this field.

To fully appreciate this, one must recognize and understand the demands medicine makes on manufacturers to develop and provide high-quality modalities that ensure users the highest level of patient care and deliver the best possible support in the areas of quality management, administration, law, education and research.

Workflow is based on a number of realistic clinical patient-care scenarios (Fig. 10.2).

- Patient registration, i.e., input of all relevant patient data
- Order entry, examination plan and scheduling, e.g., modality and order lists

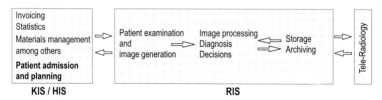

Figure 10.2 Schematic diagram of patient-oriented workflow in a hospital with a local network. The levels of function are divided into patient registration, image acquisition, image viewing and reporting, image archiving, image documentation and information exchange.

10 Patient Data Management

- Examination, report data entry and documentation, image acquisition, exam data management
- Image buffering and archiving
- Image query/retrieve, image viewing and processing
- Service data entry, generate service and supplies statistics, etc.

The structure of the local communication network in a hospital depends on the existing communication modalities and those that need enhancing, the forward-looking requirements of workflow and the associated developing system and network strategies.

10.4 Networks

Electronic data is communicated through networks. Networks, i.e., all the modalities used to exchange digital data, are connected to each other using data lines (cables). They form the foundation for ward-independent communication as well as define its geographic bounds:

- Local networks up to 5,000 m within hospitals (LAN, Local Area Network),
- Regional networks, i.e., within a geographically defined region up to 20 km (MAN, Metropolitan Area Network) and
- Interregional networks over areas 20 km and larger, with the ability to communicate worldwide (WAN, Wide Area Network).

Electronic data flowing through networks extends well beyond patient data management and the documentation and archiving of patient records. It is the medium of the future.

10.5 Local Network Structures

To handle all patient-oriented processes in the hospital, all necessary communication modalities are linked via a local network (LAN). These modalities include:

- Medical information systems such as the HIS (Hospital Information System) and RIS (Radiology Information System)
- Imaging and recording system modalities for digital radiography (DR), computer tomography (CT), magnetic resonance imaging (MRI), nuclear medicine (NM) and/or ultrasound (US)

10.5 Local Network Structures

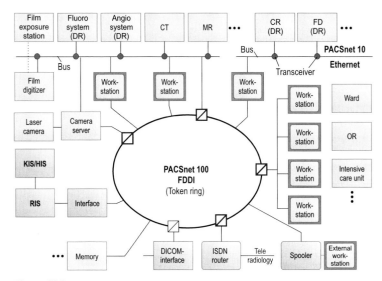

Figure 10.3
Schematic diagram of one possible hospital network (also refer to the next section)

- PACS (Picture Archiving and Communication System).

Depending on the existing system and communications potential in a hospital, different configurations for wiring imaging and information systems may have evolved into a network over time. These wiring structures are referred to as topologies. For reasons of security, reliability, speed and even cost, most networks are a mix of structures, called "heterogeneous networks."

10.5.1 Topologies

The most common topologies are based on their geometry:

Bus topology (line topology)

Bus topology is a data transmission bus with one central cable. Each workstation is attached in series, much like stops along a bus route.

Star topology

Star topology is a star-shaped configuration built around a central hub (star switch) with transmission lines that extend outward to each device. To add workstations to a star topology network, additional cables must be installed between the central hub and the device.

279

Ring topology

Ring topology is a shared closed loop to which individual nodes are attached. In turn, each node can provide service to several workstations wired to it.

Ethernet

The standard Ethernet topology uses a coaxial cable (about the thickness of a finger) with a data transmission rate of 10 Mbit/s (10 million bits per second). This Ethernet, based on the familiar "yellow cable", is the standard Ethernet /10Base-5. It is used together with a bus topology to connect system workstations and uses the CSMA/CD (Carrier Sense Multiple Access/Collision Detection) access procedure, i.e., access to the bus (bus cable) is possible only if no other workstation is attempting to send data to the bus via the transceiver (transmit and receive device). (See Figures 10.3 and 10.4.)

Token Ring

Token ring is an example of ring topology in the local network (LAN) (see Fig. 10.4). It uses a twisted-pair cable (see Fig. 10.6) with data transmission rates up to a maximum of 16 Mbit/s. In this example, the data transmission medium is set up as a closed loop. All the communication and information systems that are part of the ring topology are connected via a controller (interface). An interface is a special connection point to electronically adapt two non-compatible systems or system components. Here, the interface functions as a controller, i.e., a control switch.

Similar to the send permission of the bus topology, only one of the connected workstations can send data over the ring. To ensure that only one message or data from one workstation is sent over the network, a token circulates in the loop from controller to controller. A token is a control signal consisting of a series of bits or a byte with a specific pattern. Only the workstation whose controller "possesses" the byte (has taken it from the ring) can transmit data over the network at that time (token passing).

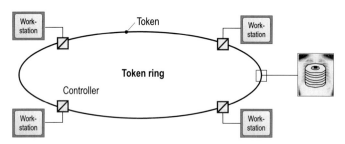

Figure 10.4 Schematic diagram of a bus topology and a token ring

FDDI network

The FDDI (Fiber Distributed Data Interface) network is a ring topology in a local network (LAN) (Fig. 10.5). It uses a fiber-optic cable or twisted pair (see Fig. 10.6) with a transmission rate at this writing of 100 Mbit/s.

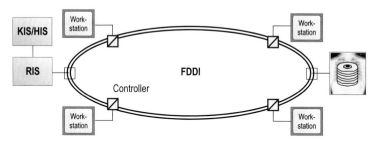

Figure 10.5 Schematic diagram of an FDDI network

Physically, the FDDI network consists of a double ring. During a normal work process, data is communicated over one ring only, the primary ring. The other ring, the secondary ring, serves as a backup for the data transfer and takes over in case the primary ring fails or has trouble. Similar to bus or token-ring topology, data-send permission is regulated via token passing.

As radiography systems become increasingly digitized, the amount of the data to manage will become enormous. Compared to twisted-pair topologies, FDDI networks are relatively expensive as well as costly to install and maintain due to their relatively complex connection technology. However, because of their high transmission rate (min. 100 Mbit/s), electromagnetic robustness and reliability, FDDI networks often are the basis of PACS installations.

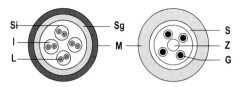

M: Outside sheath
L: Conductor
I: Insulation
Si: Inner shielding for wire pair
Sg: Outer shielding for twisted pair cable

Z: Strain relief
S: Cable core containing supporting elements of synthetics and stranding elements
G: Gradient fibers.

Figure 10.6 Cross-section of a coaxial cable (left), known as a twisted pair, shown here in "4 × 2" configuration, i.e., four twisted-wire pairs. Cross-section of fiber optic cable (right), here with stranding elements.

See the appendix for additional references and in-depth information on network standards such as the Ethernet, FDDI and ATM (Asynchronous Transfer Mode), as well as network components such as the repeater, workgroup hub, bridge, etc.

10.6 PACS

Digital technology brought enormous space savings to radiology for image archives and time-consuming communication. As the use of digital imaging systems increased, a model for archiving this image data emerged in the 1970s and was termed PACS.

- PACS, Picture Archiving and Communication Systems. PACS is a picture archiving and communication system based on a local, modular, and usually heterogeneous network (LAN).

To process and communicate all the work processes in a hospital, all the imaging system modalities, the radiology information system (RIS) and the hospital information system (HIS) are linked to one another via transceivers, controllers, concentrators, interfaces, gateways, etc.

The exchange of data and information in a hospital can be comprehensive only when the X-ray dept. PACS (as well as other specially configured networks, such as in cardiology, intensive care, the lab and OR, etc.) are closely interwoven with the Radiology Information system (RIS) and able to communicate with the Hospital Information System (HIS).

These modalities, widely varied in focus, tend to come from different manufacturers and can work in a number of data formats. To ensure standardized communication, all the data to be shared must be prepared in a standard data format. (See section 10.10.2 for information on DICOM and HL7 and section 10.10.3 for IHE.)

Figure 10.7 depicts one possible workflow in a hospital.

Patient data management usually begins on admission, when the patient's personal data is entered in or added to the Hospital Information System (HIS).

Examination results determine subsequent diagnostic procedures, e.g., a radiologic study. The modalities and schedule (modality worklist) are set using the existing patient data and the Radiologic Information System (RIS). The new result images from the study are sent to the RIS and completed with the available patient data.

10.7 Communication and Storage

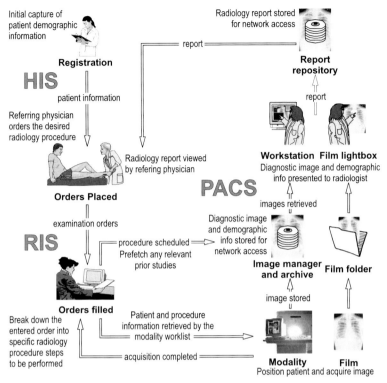

Figure 10.7
Schematic diagram of possible clinical workflow in a hospital with a radiology dept

The current images generated on the system modalities and any earlier images available or requested from the RIS are sent to the data or image data management server. From there, the data goes to the image-viewing and processing workstations for evaluation, processing and reporting.

New findings from the exam such as reports, treatment measures, etc., are buffered and sent to the HIS and completed with the available patient data. (See section 10.9 for more information on the HIS).

10.7 Communication and Storage

Once an image is generated, it is transmitted to the viewing monitors, either directly or via buffer, and to various image analysis and processing workstations. The image is then transmitted for storage or temporary storage from one archive to the next. Another normal task of the imaging workflow is retrieving the data after it is archived.

10.7.1 Image acquisition station

The images are stored on a local drive of the image acquisition system, also known as the image acquisition or imaging station. Since the patient may have only one follow-up exam, the images are stored there for one or two days.

10.7.2 Image analysis and processing workstations

Current and previously saved images, even image data generated on other modalities, can be retrieved from image analysis and processing workstations (Fig. 10.8). These workstations are, in effect, the main workstation for image optimization and processing, reporting, communication with other wards (e.g., cardiology), as well as for data storage, organization, documentation and archiving.

The reporting workstation can be set up to automatically retrieve previous studies from the archive for comparison. At the same time, the exams are automatically transmitted to the image and patient data workstation, where they are saved to RAID. Completed exam reports are archived on long-term storage media and are available in the jukeboxes for some time. Once reporting is completed and archiving confirmed, the reports are automatically deleted from the local workstation database.

As a rule, several different images are displayed, making fast access to all images, both past and present, critical. Storage capacity and access time must accommodate the amount of data. Current patient image data are stored on the local drive for approximately one week.

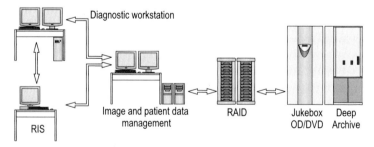

Figure 10.8
Imaging systems send digital exams to the workstation of the diagnosing radiologist

10.7.3 RAID

RAID (Redundant Array of Inexpensive Disks) is the central PACS medium for managing image data (Fig. 10.8). It is a very large, fast and reliable storage device for medium-term archiving (six to twelve months). It consists of a grouping of several inexpensive hard drives that automatically stores the data redundantly. If one drive fails, e.g., if it is destroyed, the other drives contain enough information to reconstruct the data from the defective drive.

The memory for the image data management receives image maps from connected workstations and imaging systems. Temporary memory can be accessed for images needed very quickly, such as for reporting or demonstration purposes. If the RIS is linked, old images can be prefetched automatically and used for reporting.

Data must be available quickly throughout the patient's hospitalization. The main memory controls the period that images remain in temporary memory and when they are transferred to the long-term archive.

10.7.4 Long-term archiving (jukebox)

Long-term archiving requires digital data media with an extremely high storage capacity. In a university hospital, for example, the storage requirement may be several terabytes (1 terabyte = 10^{12} bytes). Digital-optical media or tapes may be used for long-term archiving.

Optical drives or discs (OD) or magneto-optical discs (MOD) are round, plastic-coated discs. They are used for mass storage of bits. They are recorded magneto-optically, i.e., using a laser beam, causing the data to be saved as logical 1s and 0s (see digitization) as bits of differing magnetism (different magnetic direction, polarization). A laser beam is also used to read the data; the stored magnetic direction is converted into the corresponding light polarizations.

A long-term digital archive is built using 5 1/4" CDs (compact disks), which are differentiated as follows: CD-R (recordable only), CD-ROM (read-only memory), or CD-WORM (write once, read many) or DVD (digital versatile disk; DVD-R is write once, read many and DVD-RAM is write many). Access time ranges between several seconds and 2 minutes.

For long-term online archiving over several years, an automatic tape archive is usually used, called a deep archive or tape robot. Similar to video cassettes, tape access is automatic once the patient data is requested. Access time is between 2-5 minutes. Tape capacity starts at 1 terabyte (10^{12} bytes).

10.7.5 Spoolers

A radiology department usually uses equipment from a variety of manufacturers. So that the different systems can exchange data, an SPI spooler (Standard Product Interconnect) is required. (See section 10.11.1).

10.7.6 Routers

Routers connect networks of differing structures, e.g., within a hospital (LAN), within a regional network (MAN), or to wide area networks (WAN). (See section 10.4).

10.7.7 Film digitizers

So that all examinations results are available in digital form, including images generated on analog systems, the film digitizer converts analog images into digital images.

10.8 Metropolitan and Wide-area Communication

The reporting workstation and the workstation for image and patient-data management (Fig. 10.8) transmit data for the most part automatically (autorouting). The data can be limited to only those images that are relevant to the report. The web server thus reduces the number of accesses to the digital archive by the workstations and practices.

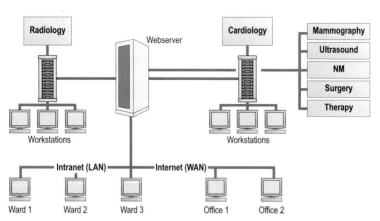

Figure 10.9 The web server receives images from the diagnostic departments using the DICOM protocol

In addition to radiologic systems, manufacturers do offer their own bus networks for special clinical functions that are networked with the PACS. They include special diagnostic or treatment functions, such as cardiology, oncology, the lab, the OR and intensive care (Fig. 10.9).

10.8.1 Wide-area networks

Radiologic images, reports and patient data can be transmitted all over the world via an ISDN connection. This form of "teleradiology" not only offers online exchange of images and data, but also communication in the form of on-call duty, reporting or expert consulting, etc., from every corner of the earth.

10.9 Hospital and Radiology Information Systems (HIS/RIS)

It is essential for a well functioning information system to provide the right data at the right time and place for any given field of endeavor. Electronic information systems are modalities that store, retrieve, correlate and analyze data. They are comprised of a data processing system, a database (a database system) and analysis programs.

In healthcare, electronic information systems are used by physician's practices and university hospitals alike.

The tasks of the information system are to secure and speed the flow of information in the hospital as well as to simplify and rationalize workflows. This includes not only medical data, but administrative data as well. The information systems record and save all the data and must be able to make it available at all times.

10.9.1 HIS

The Hospital Information System (HIS) is the hospital's main information system.

The HIS is a subsystem that comprises hardware and software to perform all data processing transactions in a hospital, including the persons (agents) who use it (entry, processing, storage, retrieval, analysis, etc.).

See the references section for in-depth coverage of the HIS, medical information science and medical documentation.

10 Patient Data Management

10.9.2 RIS

Similar to the HIS, which covers all hospital data-processing tasks, the RIS (Radiologic Information System) is a subsystem for electronically processing the data of the hospital's X-ray department. The RIS falls under diagnostic radiology and is integral to the HIS; data interchange between the HIS and RIS must be rigorous if hospital communication is to be thorough.

The RIS transmits all the required patient and image data, including image data from modalities other than X-ray. All the digital or digitized images and patient data from the imaging systems are recorded in the RIS, including data from magnetic resonance imaging and nuclear medicine.

The RIS is able to supply new or existing patient data as well as schedule and manage examinations. It is used to manage to the exam process, including the anticipated imaging modalities, the required exam and image parameters (modality worklist), and the personnel needed (workflow management).

10.10 Standards

Communication systems are based on the Healthcare Communication Standards (HCS, Fig. 10.10).

Imaging systems are based on the DICOM (Digital Imaging and Communications in Medicine) 3.0 standard of the NEMA (National Electrical Manufacturers Association).

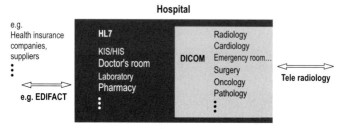

Figure 10.10 Communication standards in healthcare

Information systems are based on the Health Level 7 (HL7) standard recognized by ANSI (American National Standards Institute).

10.10.1 ACR/NEMA 2.0

Any given hospital radiology department uses image modalities from a variety of manufacturers.

To enable the digital transfer of image data between the various systems, the American College of Radiology (ACR) and NEMA introduced the ACR/NEMA 2.0 standard. It defines the individual parameters of the digital image data based on the electronic communication level. This PACS-compliant ACR/NEMA standard is known as the Standard Product Interconnect (SPI).

10.10.2 DICOM 3.0

The array of different imaging systems and manufacturers gave rise to numerous digital image formats. In 1982, ACR and NEMA co-founded a committee to develop a standard with guidelines for the formatting and exchange of images and related data. Thus, DICOM was born. To this day, DICOM 3.0 is the current industry standard for the communication of medical images and textual data. It is a consistent extension of ACR/NEMA 2.0 and contains network-wide services and standardized data formats.

For exchanging digital images from diagnostic imaging modalities, DICOM uses the standard network connections and protocols of the computer industry. DICOM specifies the structure of the image formats and descriptive parameters for images or image series and commands for data exchange.

The DICOM standard is limited to medical imaging systems, such as diagnostic X-ray procedures, digitized X-ray images, fluoroscopy, computed tomography, nuclear medicine, ultrasound, magnetic resonance imaging, as well as image-viewing and processing workstations, HIS/RIS connections and network-capable hardcopy systems (e.g., laser printers).

DICOM Conformance Statement

The DICOM Conformance Statement is a document made up of multiple parts. It describes in detail how the modalities exchange commands and data and how information objects such as graphics, images, studies, reports, etc., travel through the network.

The terms DICOM compliant/DICOM compatible are used to identify systems and modalities that are able to perform certain functions in conformance with the DICOM standard.

Service classes specify the level of accuracy. To indicate that a product is DICOM compatible is not enough. To guarantee complete network capability requires a detailed classification.

The functionality of the DICOM interfaces as described in the DICOM Conformance Statement is mandatory. Overall interface functions with or between different systems require explicit validation, i.e., a statement of reliability of the functional interface and its release.

10 Patient Data Management

Using DICOM 3.0, every image generated from a DICOM-compatible system from any manufacturer and transmitted in a DICOM network can be found, loaded, displayed, printed, saved and or archived.

The following lists the global interface functions used most frequently:

- DICOM Send
  Transmits image data from one modality to the network (network interface with integrated DICOM SEND in the DICOM 3 standard).
- DICOM Print
  Transmits image data from one modality over a network to a DICOM-compatible documentation/hardcopy system (e.g., a laser or network printer).
- DICOM Query/Retrieve
  Query: Searches for images (image data) for a specific patient on the network.
  Retrieve: Transmits images (image data) to the selected image evaluation and processing workstation when the system has located the images via Query (provides DICOM 3 Query/Retrieve services for connecting to a patient image archive).
- DICOM MPPS/SC (StC)
  MPPS = Modality Performed Procedure Step: Sends reports to a patient data management system (RIS). SC (StC), Storage Commitment: Confirms archiving from the image archive (RIS).
- DICOM Get Worklist or DICOM Modality Worklist
  Imports patient data and image requirements from the patient management and information systems (HIS/RIS).

10.10.3 Health Level 7 (HL7)

Hospital information systems usually use the Health Level 7 (HL7, Fig. 10.11) communication standard to transmit patient-related data (without image data). The systems that use HL7 include the HIS, lab, pharmacy, physicians' offices, etc.

	HIS	RIS	PACS	Modality
HIS	HL7	HL7	—	—
RIS	HL7	HL7	HL7/DICOM	DICOM
PACS	—	HL7/DICOM	DICOM	DICOM
Modality	—	DICOM	DICOM	DICOM

Figure 10.11 Communication standards used for transmitting data between information systems, PACS and imaging systems (modalities)

10.10.4 EDIFACT

EDIFACT (Electronic Data Interchange for Administration, Commerce and Transport) is a communication standard of the United Nations (UN) (See Fig. 10.10). Yet EDIFACT has not succeeded in replacing HL7 as the standard inside hospital information systems.

EDIFACT remains in use, however, for hospital communications with its business partners, such as suppliers (orders, delivery slips, invoices, etc.) or health insurers (transmission of diagnosis and treatment data).

10.11 GIF, JPEG, TIFF Image Format Standards

Several different simple image formats (standards) have been established for saving image data. Simple bitmap formats like GIF or TIFF are not suitable for the electronic transfer of data within telemedicine.

10.11.1 GIF, Graphics Interchange Format

GIF images (.gif file suffix) contain a maximum of 256 colors, whereby a specific color can be defined as the transparent background. GIF images can be compressed without any data loss (known as lossless compression). The quality is not compromised; any redundant image information is simply not stored multiple times, so the compression rate is not very high. For example, the larger the areas of identical color are, the more effective this method of compression is.

10.11.2 JPEG, Joint Photographic Expert Group

In the JPEG format (.jpg file suffix), images with a high number of colors are compressed much more tightly. During compression, however, more or less image data are lost and cannot be recovered (known as lossy compression). When images compressed using the JPEG format are enlarged, their color contours are blurred.

In general, the larger the image (graphic), the more likely it is that GIF will produce a smaller file. The more color detail the image has, the more likely it is to be saved in JPEG format.

10.11.3 TIFF, Tagged Image File Format

TIFF format (.tif file suffix), like GIF, retains all data during compression (lossless compression). TIFF has established itself as simple, all-round compatible file format, although it is not particularly suited for display over the internet.

10.12 Integrating the Healthcare Enterprise (IHE)

The IHE is an international standards-based industry consortium founded in late 1998 by the RSNA (Radiological Society of North America) and the HIMSS (Hospital Information and Management System Society). The organization's goal is to integrate systems from different manufacturers and optimize clinical workflows by applying proven IT standards in healthcare. Objectives include to:

- Further communication between the different suppliers of medical information technologies.
- Support the application of communication standards such as HL7 and DICOM in healthcare.
- Expand access to clinical information. Continuity of patient data.
- Close the data loop of the various medical information systems such as the HIS, RIS, PACs, imaging and recording system modalities, to eliminate redundant data and repetition of work steps.

The IHE model is based solely on standards that are already in existence. The general framework of the IHE unites the traditional world of information systems, based on HL7, and the world of imaging systems, based on DICOM, into a single integrated environment. For more information on the IHE, consult the following website:

- http://www.siemensmedical.com/ihe

11 Appendix

11.1 Glossary of Terms

The following concepts and terminology in medicine and technology are provided as background information and are listed in alphabetical order:

Ampere

André Marie Ampère, French physicist and mathematician (1775 to 1836), is one of the founders of electrodynamics. The base unit of electric current was named after him. The internationally recognized ampere is defined as the constant current that flows through two straight parallel conductors one meter apart and produces between the conductors a force of $2 \cdot 10^{-7}$ newton per meter of wire. The current through the wire is equal to 1 A (ampere) if a charge of one coulomb per second flows through its circular cross-section.

Anatomy

The study of the structural makeup of the human body.

Angiocardiography

See cardiac angiography.

Angiography

The radiographic visualization of blood vessels after injection of a contrast agent in a vessel.

Aortography

Visualization of the aorta (and its arteries) after injection of a contrast agent after direct puncture of the aorta or insertion of a catheter from a leg or arm artery.

Arteriography

Visualization of arteries in a specific part of the body after injection of a contrast agent in a larger artery supplying the region by means of vascular system interventions, arterial, i.e., translucent, oxygenated blood (general angiography).

Arthrography

Contrast visualization of joints after injection of positive and negative contrast agents in the joint cavity; also as a double contrast visualization.

Atom

Atom, from Gk. "atomos," indivisible, the smallest particle of a chemical element. Atoms are indestructible by chemical means but they can be split using physical methods.

Atomic model

Ernest Rutherford, British physical chemist, 1871 to 1937 (1908 Nobel Prize in chemistry), discovered and studied the radioactive decay of elements and divided emitted radiation into α-, β- and γ-beams. Based on his observations of the scatter of α particles at atomic nuclei, he developed the "Rutherfordian atomic model," in which he describes the sun as the nucleus and the planets as electrons (planetary model).

This model was further developed by Niels Bohr, Danish Physicist, 1885 to 1962 (1922 Nobel Prize in physics) and Arnold Sommerfeld, German physicist and mathematician, 1868 to 1951. N. Bohr applied principles from quantum theory to Rutherford's model. The result is the eponymous "Bohr's atomic model," in which electrons circle the atomic nucleus on paths known as "shells." Bohr's model serves as the foundation for modern quantum theory.

Arnold Sommerfeld expanded this model in the "Sommerfeld atomic model" by establishing that electrons also move in elliptical paths at the same energy level.

Atomic number

The atomic number Z is the number of positively charged particles (protons) in the nucleus of the atom. Z defines the position of a chemical element in the periodic table of elements. All known elements are listed in the periodic table of elements according to their atomic mass (formerly known as atomic weight).

Simultaneously in 1869 (but independently of one another), Dmitri Ivanowitsch Mendeleyev (Russian chemist, 1834-1907) and Lothar Mayer (German chemist, 1830-1895) arranged the chemical elements into rows (periods) in order of their atomic mass and further organized them into groups.

Brachial angiography

Visualization of arteries after injection of a positive contrast agent in the upper arm artery, including countercurrent angiography for visualization of arm and hand arteries.

Bronchography

Visualization of the bronchial system after injection of a positive contrast agent with the help of catheter guided through the mouth or nose (through the larynx in the trachea and bronchial tubes).

Bucky

Gustav Bucky, 1880 to 1963, radiologist in Berlin and New York, recognized the unfavorable effect of scatter radiation on image quality; in 1912 he developed the honeycomb diaphragm, a grid consisting of longitudinal and transverse lines; his idea for the scatter radiation grid remains indispensable even today.

Candela

Candela (Latin for candle) was introduced in 1948 as an international unit of light intensity. Based on a decision by the General Conference for SI units (SI = Système International d'Unités): 1 cd is the candle power in a specific direction of a radiation source, which emits monochromatic radiation of $540 \cdot 10^{12}$ Hz or a wavelength of 555 nm and whose intensity is 1/673 Watt per steradian (solid angle); the luminance is the light intensity of a light source whose light is being emitted from a specific surface area (cd/m^2).

Cardiac angiography

Visualization of the interior of the heart and the large neighboring arteries after injection of a positive contrast agent in one of the cavities of the heart with a catheter (inserted through an arm or leg vessel).

Celiacography

Visualization of the truncus coeliacus, the artery that supplies upper abdominal organs (liver, stomach, spleen and pancreas) after injection of a positive contrast agent through a catheter inserted into it (through a vessel in the leg).

Cerebral angiography

Visualization of arteries and veins of the brain after injection of a positive contrast agent into the carotid artery or into the aortic arch.

Cholangiography

Visualization of the bile ducts after intravenous injection of a positive contrast agent or after draining the gall bladder filled with a contrast agent (cf. cholecystography). Includes operative cholangiography after injection of a positive contrast agent in the bile ducts and postoperative cholangiography

11 Appendix

after injection of a positive contrast agent in a T-drain inserted in the bile ducts during the operation.

Cholecystography

Visualization of the gall bladder after oral administration or intravenous injection of a positive contrast agent (oral or intravenous cholecystography).

Cisternography

Visualization of the basal cisterns of the base of the brain with a negative contrast agent.

Colon contrast enema

Visualization of the large intestine after instillation of a positive unmetabolized contrast agent in the anus through an intestinal tube, often a double-contrast study.

Comberg technique

Procedure for localizing foreign objects in the eye (orbital).

Compton

Arthur Holly Compton, American physicist, 1892 to 1962 (1927 Nobel Prize for the discovery of the Compton effect). A.H. Compton also discovered a technique for directly determining the wavelengths of X-rays using an optical grid.

Computed Tomography (CT)

Method of X-ray examination using cross-sectional scans along a single axis.

Contrast agents

Substances that have a higher atomic number (positive contrast agent) or a lower atomic number (negative contrast agent) than the surrounding bodily tissue. Contrast agents allow the display of organs that normally cannot be seen or are difficult to see on plain film radiographs.

For example, barium-sulfate-based agents are used for visualization of the gastrointestinal tract, and iodine-based agents for the blood vessels. Negative contrast agents include air or other gases. Double-contrast methods combine positive and negative contrast agents.

Coronary angiography

Visualization of the coronary arteries after injection of a positive contrast agent through a catheter inserted in the mouth of the artery (at the aorta). Also general coronary angiography after injection of a positive contrast agent through a specially shaped catheter at the start of the aorta near the heart.

Coulomb

Charles Augustin de Coulomb, French engineer and physicist (1736 to 1806). Coulomb's Law states that the force F between two electric charges Q_1 and Q_2 at a distance of r is $\sim Q_1 \cdot Q_2/r^2$. Charges of the same sign (+/–) are repulsed, otherwise they are attracted. 1 coulomb (C) is the electric charge transferred by a current of 1 ampere in 1 second (As).

Countercurrent angiography

Visualization of vascular sections upstream (i.e., towards the heart for arteries) from the injection site or branching off from a larger vessel after injection of a positive contrast agent in a blood vessel (usually an artery) against the flow using an injection pressure greater than the blood pressure in the vessel.

Cystography

Visualization of the urinary bladder with a positive contrast agent, either after filling the bladder or in the course of an excretory urography.

Decibel

Decibel: deci-, tenth; bel, after the US physiologist and inventor A.G. Bell (1847 to 1922). Bel is a key concept for units of measurement and represents the ratio of two amounts of electric or acoustic signal power equal to 10 times the common logarithm of this ratio (see chapter 8 for information on X-ray film and optical density).

Density

The density (rho) is the ratio of the mass m of a substance (matter) in grams (g) to its volume V (cm^3): $\rho = m/V$ (g/cm^3).

Detective quantum efficiency

See DQE

11 Appendix

Digital fluoro radiography (DFR)

Digital radiography with an image intensifier or flat detector as the image receptor.

Digital luminescence radiography (DLR)

Digital radiography with imaging plates (storage phosphors) as the image receptor (e.g., computed tomography).

Dilatation

The action of dilating or enlarging a vessel.

Double contrast methods

Visualization of hollow spaces (e.g., gastrointestinal tract, joint cavities) after lining the relevant organ using a positive contrast agent and then inflating it with a negative contrast agent, usually air.

DQE

Detective quantum efficiency or quantum strength (e.g., noise equivalent quanta, quantum detector response, effective quantum utilization) represents the percentage of quanta used by the detector to create an image.

The smaller percentage of quanta striking the detectors of the image receptor system penetrates the input screen; the larger percentage is absorbed during the imaging process (luminescence). The DQE is the percentage of X-ray quanta actually detected by the receptor.

Embolization

Process by which a blood vessel or organ is obstructed by a material mass.

Encephalography

See pneumoencephalography.

Endoscopy

Visualization of the interior of a hollow organ using an endoscope.

Enlargement image

An enlargement is produced with a very small focal point, e.g., 0.3 to 0.1. The enlargement is a result of the distance ratios between the focal point, the object and image receptor plane.

Esophagram

Visualization of the esophagus by swallowing a positive unmetabolized contrast agent (cf. gastrointestinal tract).

ESWL

Extracorporal shock wave lithotripsy: disintegration of stones through concentrated, high-energy sound waves (shock waves) generated outside the body. X-ray radiation or ultrasound is used to localize the stones.

Excretory urography

Visualization of kidneys, pelvis, ureters and bladder after injection of a positive contrast agent (cf., pyelography, urography).

Femoral angiography

Visualization of leg arteries after injection of a positive contrast agent in the femoral artery (Arteria femoralis).

Fistulography

Visualization of fistulas leading to external body surfaces after instillation of a positive contrast agent in the mouth of the fistula.

Galactography

Visualization of the lactiferous ducts after injection of a positive contrast agent in the opening of the ducts being examined.

Gastrointestinal tract

Visualization of the gastrointestinal tract after drinking of a positive, unmetabolized contrast agent (in special cases, a resorbable positive contrast agent is used). Double-contrast imaging of the stomach or esophagus (using its natural air content or insufflation of air). Double-contrast imaging of the small intestine using the Sellink method with a positive contrast agent and cellulose as a negative contrast agent, intubated via an intestinal probe.

Glandulography

Visualization of the duct systems in glands (e.g., salivary gland) after instillation of a positive contrast agent in the excretory duct of the gland.

Gray

Gy = Gray, since 1986 the only approved unit of energy dose; named after the US physicist Louis Harold Gray (1905-1965) for his major contributions in radiation dosimetry; 1 Gy = 1 J/kg = 100 rad (former designation).

11 Appendix

Halide

Halogen is the group name of chemical elements that form salts (crystals) with metals in the absence of oxygen. Halides belong to group 17 of the periodic table. Fluorine (atomic symbol F, atomic number 9), Chlorine (Cl, 17), Bromine (Br, 35), Iodine (I, 53) and Astatine (At, 85). Halide is the group name for the anorganic and organic bonding of halogens.

Heat Units

Heat units (HU) are not to be confused with Hounsfield units (HU): the Hounsfield unit is used in CT as a dimension for measuring the attenuation value of a CT image (e.g., -1000 HU for air, 0 HU for water, up to 2000 HU for bone). In CT, so-called CT numbers are used, relative to the attenuation of water.

Hepatography

Visualization of the liver after uniform opacification with contrast agent; used today primarily during angiography of the hepatic arteries or a splenoportography.

Hertz

Heinrich Hertz (1857 to 1894), German physicist, verified in 1887/88 the predictions of J.C. Maxwell regarding the coexistence of long electromagnetic waves (radio waves) with light waves. The unit of frequency or hertz (Hz) was named after him. 1 Hz = 1 one cycle per second: 1 kHz (kilohertz) = 10^3 Hz; 1 MHz (megahertz) = 10^6 Hz, etc.

Hysterosalpingography

Visualization of the hollow spaces of the uterus and fallopian tubes after installation of positive contrast agent in the cervix by means of a special instrument. The technique is used to check the patency of the connection between the fallopian tubes and free abdominal cavity.

Infusion urography

Visualization of kidneys and urinary tract after injection of a diluted, positive renal contrast agent in a vein.

Intensity I_ψ

The intensity I_ψ, or rather the energy flux density of X-rays, is the radiation field restricted to a specific energy (sum of all X-ray quanta, photons) that travels in one second through 1 cm^2 perpendicular to the propagation direction of the radiation. The intensity depends on the tube voltage U_R and the anode material.

Interventional radiography

Diagnoses and subsequent therapy supported by various percutaneous (through the skin) interventions under fluoroscopy.

Joule

James Prescott Joule (1818 to 1889), British physicist; defined the mechanical heat equivalent; the energy unit named after him is the joule (1 J = 1 Ws, work done in a specific time period). The energy E = 1 eV = $1.602 \cdot 10^{-19}$ J ($1.602 \cdot 10^{-19}$ J is the elementary charge of an electron).

Kerma

Dose quantity. Acronym for "Kinetic energy released in matter." Kerma is the ratio of the sum of the initial kinetic energies (dE) for all charged particles that are released through photons and indirect ionizing radiation in a volume element of the irradiated material with the mass dm. The kerma must always specify the material, e.g., air kerma. Unit: 1 Joule/kg = 1 Gray (Gy).

Kidney angiography

Visualization of the arteries and veins of the kidneys (including temporarily staining the kidneys using a contrast agent) by injecting a positive contrast agent into the arteries of the kidney via a catheter puncturing either the aorta close to the kidney arteries (general angiography) or directly into the kidney artery (selective kidney angiography).

Kymography

Obsolete imaging technique: X-ray images of independently moving organs (heart, stomach, intestines) with a lead slotted grid positioned between the object and the film. Either the film is moved with the grid in place or the grid is moved with the film in place.

Laser

Generator of monochromatic nearly parallel beams of light through light amplification by stimulated emission of radiation. Gas or semi-conductor lasers. High energy densities can be achieved.

Logarithm

Logarithms are considered the number (base 10, in this case), to whose power another number (the base, here I_0 and I_1) must be raised to obtain a predefined number (the number for which the logarithm is sought): For $10^1 =$ 10, the base-10 logarithm of 10 is 1; for $10^2 = 100$, the base-10 logarithm of 100 is 2.

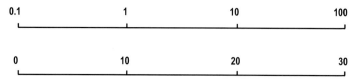

Figure 11.1 Comparison of a linear (bottom) and logarithmic measurement (top): In order to show the linear measure in the 100 range, the abscissa needs to be extended to the far right.

Lymphography

Visualization of lymph vessels and nodes after injection of a positive contrast agent into a lymph vessel (made visible and exposed by a special dye) usually on the dorsum of the foot or sometimes the hand, infrequently on other parts of the body.

Mammography

Radiographic examination of the female breast.

Mass

Mass causes inertia (resistance) and all bodies to have weight (in gravity).

Mediastinography

Visualization of the mediastinum (area between the lungs) after injection of a negative contrast agent (cf. pneumomediastinum, pneumoretroperitoneum).

Micturating urethography

Visualization of the urethra during micturation after previously filling the urinary bladder with positive contrast agent (using excretory urography or a catheter inserted into the urinary bladder).

Myelography

Visualization of the subarachnoid space (between the arachnoid membrane and the spinal cord membrane) in the region of the spinal cord after injection of a positive contrast agent by puncturing the lumbar spinal canal (spinal tap) or the craniocervical junction (subocciptal puncture).

Nephrography

Visualization of the kidneys after staining with a positive contrast agent during excretory urography, infusion urography or kidney angiography.

Percutaneous transluminal cholangiography (PTC)

X-ray contrast visualization of the biliary ducts and the gall bladder.

Percutaneous transluminal coronary angioplasty (PTCA)

Dilatation, e.g., using a balloon catheter.

Peripheral angiography

Visualization of the blood vessels of the legs or arms after injection of a positive contrast agent into an artery of the arm or leg (cf. brachial angiography, femoral angiography).

Pharmacoangiography

Visualization of blood vessels after injection of a positive contrast agent as in selective angiography after preliminary injection of a vasoconstrictive or vasodilating substance through the same catheter.

Pharmacoradiography

Visualization of the various organ systems, especially the gastrointestinal tract, with contrast agents after administration of drugs that affect the tonicity (tonus) and the autonomous movement of the organ under study.

Phlebography

See venography.

Planck's constant

Quantum energy $E = h \cdot v$, where h is Planck's constant and v is the frequency of the quantum radiation.

Max Planck, German physicist (1858-1947), founder of quantum theory, Nobel Prize for physics in 1918. The study of the radiation of heated bodies led him to discover that energy is not emitted continuously, but emitted in tiny portions, known as quanta.

In 1900 he discovered the natural constant h with the unit of work: h = $6.62559 \cdot 10^{-34}$ J·s (joule × seconds). The product of work or energy and time, impulse and length, angular momentum and angle is known in physics as a unit of work.

Planigraphy

X-ray slice image by movement of the X-ray tube and film in parallel planes (cf. tomography).

Pneumarthrography

Visualization of interior of a joint after injection with a negative contrast agent (cf. arthrography).

Pneumencephalography

Visualization of the fluid-containing structures of the brain and surroundings after injection of a negative contrast agent into the spinal cord.

Pneumography

Visualization using negative contrast agents (general term).

Pneumomediastinum

Visualization of mediastinal cavity after injection with negative contrast agents (cf. mediastinography).

Pneumoperitoneum

Visualization of the contours of the organs in the abdominal cavity after injection of a negative contrast agent into the abdominal cavity.

Pneumoretreperitoneum

Visualization of the connective tissue cavities behind the peritoneum including the organs inside (kidneys, adrenal glands) after injection of a negative contrast agent into the connective tissue in front of the coccyx (also retropneumoperitoneum).

Portography

Visualization of the portal vein after injection of a positive contrast agent. The term is usually used for intraoperative portography.

Pulmonary angiography

Visualization of the lungs after injection of a positive contrast agent in an arterial branch of the lungs using a catheter guided (from a vein in the arm or leg) through the right side of the heart into the lung artery.

Pyelography

Visualization of the renal pelvis, ureters and bladder with a positive contrast agent:
a) intravenous pyelography (preferred: excretory urography) by intravenous injection;
b) retrograde pyelography by injection of a contrast agent in the ureters or renal pelvis.

Reflux cystography

Visualization of the bladder by filling it with positive contrast agent through a catheter inserted through the urethra to test the reflux of contrast agent into the ureters.

Relief view

Relief view is similar to the subtraction procedure. When the original is copied with the negative/positive reverse copy or when both images are viewed atop one another on film viewer. This creates a relief image that displays the contours of the image elements more clearly.

Renovasography

See kidney angiography.

Retropneumoperitoneum

See pneumoretroperitoneum.

Salpingography

Visualization of the fallopian tubes using positive contrast agent (cf. hysterosalpingography).

Schwarzschild

Karl Schwarzschild, German physicist and astronomer (1873-1916) demonstrated that the blackening of photographic slices is not linearly dependent on the exposure (intensity times time): it starts at a certain minimum exposure and becomes saturated at a high exposure.

Selective angiography

Visualization of arteries and veins after injection of a positive contrast agent directly into the blood vessel through a catheter guided into the vessel under study through an arm or leg vessel.

SI units

SI = Système International d'Unités. International system of units. Global standard for base units of measure (length, volume, mass, etc.)

Sialography

Visualization of the salivary glands after injection of a positive contrast agent into their excretory ducts.

Sievert

Symbol Sv, after the Swedish physicist Rolf Maximilian Sievert (1896-1966), known unit of dose equivalent; in 1986, replaced the "rem": 1 Sv = 1 J/kg = 100 rem.

Simultaneous slice acquisition

Obsolete imaging technique: simultaneous acquisition of several (usually five) parallel body slices using a simultaneous slice cassette.

Slice image

See planigraphy, tomography.

Splenoportography (indirect)

Visualization of the splenic veins and the sections of the portal vein system after injection of a positive contrast agent into the superior mesenteric artery.

Stereoscopy

Obsolete imaging technique: preparation of two X-ray images at a specific distance. When the images are viewed, e.g., with a stereoscope, the impression of depth is created. The object must remain stationary between the first and second images, which are best taken one right after the other.

Thoracic aortography

Visualization of the thoracic section of the aorta and the arteries originating from it by injection of a positive contrast agent through a catheter inserted through a leg vessel into that section of the aorta.

Tomography

General term for slice acquisition procedure. Tomography differs from planigraphy in that the X-ray tube and image receptor move in circular arcs (constant SID).

Trendelenburg position

Head-down position on inclinable X-ray systems.

Urethrography

Visualization of the urethra using a positive contrast agent
a) by direct instillation of contrast agent into the urethra,
b) during urination after filling the bladder with contrast agent (micturating urethrography).

Urography

Visualization of the kidneys, renal pelvis, ureters and bladder with a positive contrast agent (cf. excretory urography).

Vasography

Visualization of blood vessels with a positive contrast agent (cf. angiography).

Wehnelt electrode

Arthur Rudolph Wehnelt, physicist 1871 (Rio de Janeiro) to 1944 (Berlin), invented the cylinder-shaped electrode that bears his name used to bundle streams of electrons.

Venography

Visualization of veins in a particular part of the body after injection of a contrast agent into a superficial vein or a section of bone peripheral to the area of interest (transosseous venography). Venous interventions (veins are carrying dark blood to the heart).

Ventriculography

Visualization of the left ventricle of the heart after injection with a contrast agent via a catheter inserted through an artery in the arm or leg.

Vertebral angiography

Visualization of the blood vessels of the cerebellum and posterior sections of the cerebrum after injection of a positive contrast agent into the vertebral artery in the neck (arteria vertebralis), either using a catheter inserted into the artery through a leg vessel or via countercurrent angiography.

Vesiculography

Visualization of the seminal vesicles after injection of a positive contrast agent into the spermatic cord.

11 Appendix

Visceral angiography

Visualization of the blood vessels of the abdominal organs after injection of a positive contrast agent into the aorta or individual branches in the aorta in the abdomen (cf. aortography, selective angiography).

Zonography

A form of tomography with a small blurring angle (approx. $8°$ to $10°$ for linear movement, $5.5°$ for circular movement). The smaller the slice angle, the thicker the slice. Application: visualization of the kidneys (cf. planigraphy).

11.2 Abbreviations

ACR	American College of Radiology
ARI	Access to Radiology Information
ANSI	American National Standards Institute
ASCII	American Standard Code for Information Interchange
CAT	Continuous Accelerated Tube
CC	Cranio-caudal
CCD	Charge-coupled Device
CCIR	Comité Consultatif International des Radiocommunications; responsible for recommending and promoting regulations for the use of common parameters for transfer and reception of TV images.
CPU	Central Processing Unit
CR	Computed Radiography
CRT	Cathode Ray Tube
CT	Computed Tomography
DICOM	Digital Imaging and Communications in Medicine
DIN	Deutsches Institut für Normung (German Institute for Standardization)
DLR	Digital Luminescence Radiography (phosphor storage plates)
DQE	Detective Quantum Efficiency
ECOG	European Community Oncology Group

11.2 Abbreviations

EMR	Electronic Medical Record
FDA	Food and Drug Administration
HIMSS	Hospital Information and Management System Society
HIS	Hospital Information System
HL7	Health Level 7
DICOM	Digital Image Communication in Medicine
DIN	German Industrial Norm (Deutsches Institut für Normung e.V.)
IEC	International Electrotechnical Commission
IEEE	Institute of Electrical and Electronic Engineers
ICRP	International Commission on Radiological Protection
ICRU	International Commission on Radiological Units and Measurements
IHE	Integrating the Healthcare Enterprise
ISDN	Integrated Serviced Digital Network
ISO	International Organization for Standardization
IT	Information Technology
JFR	Journées Francaises de Radiologie
HIS	Hospital Information System
LCD	Liquid Crystal Display
LED	Light Emitting Diode
MLO	Medio Lateral Oblique
MPPS	Modality Performed Procedure Step
MR(I)	Magnetic Resonance (Imaging)
MWL	Modality Worklist
PACS	Picture Archiving and Communication Systems
PIR	Patient Information Reconciliation
RAM	Random Access Memory
RIS	Radiology Information System
RMS	Radiology Management System
RSNA	Radiological Society of North America

11 Appendix

SINR	Simple Image and Numeric Reports
SIRM	Società Italiana di Radiologia Medica
SWF	Scheduled Workflow
TFD	Thin Film Diode
TFT	Thin Film Transistor
US	Ultrasound
WHO	World Health Organization

11.3 References

Krestel, E. (ed.): "Imaging Systems for Medical Diagnostics"; Siemens AG, 1990; ISBN 3800915642.

Morneburg, H.: "Bildgebende Systeme für die medizinische Diagnostik"; Publicis, 3rd edition 1995; ISBN 3895780022.

Kalender, W.A.: "Computed Tomography"; Publicis MCD Verlag, 2001; ISBN 3895780812.

Ewen, K.: "Moderne Bildgebung"; Georg Thieme, 1997; ISBN 3131088613.

Kamke, D., Walcher, W.: "Physik für Mediziner"; B.G. Teubner, 2nd edition 1994; ISBN 3519130483.

Pschyrembel, W.: "Pschyrembel Klinisches Wörterbuch"; Walter de Gruyter Inc., 1990, ISBN 311010881X.

Kauffmann, G. W.; Moser, E.; Sauer, R.: "Radiologie"; Urban und Fischer, 2nd edition 2001; ISBN 3437419900.

Krieger, H.: "Strahlenphysik, Dosimetrie und Strahlenschutz"; B. G. Teubner, 3rd edition 2001; ISBN: 3-519-23078-X.

Laubenberger, Th.; Laubenberger, J.: "Technik der medizinischen Radiologie", Deutscher Ärzte-Verlag, 6th edition, 1994; ISBN 3769111117.

Journals

electromedica 67 (1999) No. 2: "Healthcare Enterprise Integration" by N. Wirsz.

electromedica 70 (2002) No. 2: "IHE is Reality at Siemens Medical Solutions" by N. Wirsz, G. Oyntzen, E. Motzkus, B. Stewart.

electromedica 71 (2003) No. 1: "Integrating the Healthcare Enterprise is Reality at Siemens Medical Solutions (Part II)" by N. Wirsz, G. Oyntzen, E. Hinzmann, M. von Roden, D. Karchner, H. Primo.

Der Radiologe, Vol. 43, No. 5 (May 2003); Springer: "Flat-panel detectors in Radiological Diagnostics" by M. Spahn, R. Freytag.

Der Radiologe, Vol. 43, No. 5 (May 2003); Springer: "Digitale Thoraxradiographie Flat-panel-Detektor oder Speicherfolie?" by C. Schaefer-Prokop, M. Uffmann, J. Sailer, N. Kapalan, C. Herold, M. Prokop.

Der Radiologe, Vol. 43, No. 5 (May 2003); Springer: "Neue CR-Technologien für die digitale Radiographie": R. Fasbender, R. Schaetzing.

Radiologie up2date, Vol. 2, issue 1, Georg Thieme (www.thieme-connect.com/ejournals): "Der Einsatz von Flachbilddetektoren für die CT-Bildgebung" by W.A. Kalender.

Radiologie up2date, Vol. 2, issue 1, Georg Thieme (www.thieme-connect.com/ejournals): "Digitale Flachdetektorsysteme": S.K. Ludwig, T.M. Bernhardt.

11.4 Photo credits

All systems-related illustrations are the property of Siemens AG, Medical Solutions, unless indicated below:

Fig. 5.7:   Dr. Goos-Suprema GmbH, Heidelberg, Germany

Fig. 6.2:   SONOLINE by Siemens Medical Solutions,
Erlangen, Germany

Fig. 6.14:   Sirona Dental Systems (center, right)
by Siemens Medical Solutions, Erlangen, Germany

Fig. 8.34:   Planilux Gerätebau Felix Schulte GmbH & Co.KG.,
Warstein, Germany

Fig. 9.62-A: Liebel-Flarsheim Inc, Schering AG

Index

3D reconstruction 234

A

absorber leaves 257
absorption 35
absorption edge 42, 44
accessories 129
Al equivalent 49
angiography 93
anode 116
anode angle 120, 254
anticathode 112
area of exposure 212
arithmetic averaging 228
artifacts 246
astigmatism 164
atomic envelopes 22
attenuation coefficient 37
attenuation law 36
Auger effect 28
automap 242
automatic dose rate control 195
automatic exposure controls 191
auto-shutter 241
auto-window 241

B

bandwidth 167
basic veil 142
beam cone 32
binary code 221
binary number 222
binding energy 22
biological effect 33
biopsy, stereotactic 269

bit depth 221
bit, information unit 221
blackening degrees 140
blackening steps 144
body dose 60
bolus chase 232
bremsstrahlung 25
bremsstrahlung spectrum 30
bucky trays 126
bus topology 279
byte 223

C

cardiac angiography 94
careposition 243
careprofile 243
C-arm fluoroscopy systems 89
cassette tray 125-126
cassette unsharpness 135
catapult bucky trays 126
cathode 21, 114
CCD sensors 159, 169
characteristic curve 139
characteristic X-rays 26
circular grids 258
communication standards 288
compression 261
Compton effect 36
continuous fluoroscopy 227
contrast 248
contrast agent injectors 264
contrast agents 262

contrast index 141
contrast transfer factor 248
conversion factor 163
crossover effect 150
cross-sectional scan 269
CRT monitor 180
CT angiography (CTA) 103

D

data compression 224
data reduction 224
deep archive 285
degree of sensitivity of X-ray film 143
densitometers 142
density 22
density latitude 142
dental X-ray systems 86
detail resolution 249
detective quantum efficiency 178
Detective Quantum Efficiency (DQE) 147
DICOM, Digital Imaging and Communications in Medicine 288-289
digital cine mode 231
digital image processing 234
digital image shifting 241
digital imaging 218
Digital Subtraction Angiography (DSA) 229

Index

digitizer 286
direct flat detector systems 133
direct imaging 133
distance law 45
distortion 164
dose 55, 57
dose area product 62
dose rate 39, 61
dose-rate characteristic 196
dosimeter 69
double contrast 264
double oblique projections 265
double-slot diaphragms 124
DQE (Detective Quantum Efficiency) 147
dynamic density optimization 241
dynamic range 175
dynavision 233

E

ECG triggering 231
echomap 242
edge enhancement 238
edge filter 44
EDIFACT, Electronic Data Interchange for Administration, Commerce and Transport 291
effective dose 57
efficiency 32
electron volt 24
electronic focal spot 119
electron-optical magnification 162
electrons 22
electroradiography 153
emulsion 138
endomap 242
energy dose 58
energy flux density 30
equivalent dose 59
equivalent dose of the skin 60
Ethernet 280
examinations using contrast media 77
exposure tables 205
extrafocal radiation 121

F

FDDI network 281
film and screen unsharpness 247
film changer technology 136
film dosimeters 70
film viewing systems 179
film-screen systems 134
filters 48
finger ring dosimeter 71
flat detector technology, direct digital 176
flat detector technology, indirect digital 176
flat detectors 171, 177
fluorescent X-ray radiation 35
fluoro loop 226
fluoroscopy 190
fluoroscopy characteristic 196
flying-spot principle 155
focus jump 200
focus unsharpness 247
focusing 260
focusing cup 114
focusing range 257

G

generator output 186
generator, operating range 185
generators, continuous load 189
generators, initial load 189
generators, point technique 188
geometric unsharpness 247
GIF, Graphics Interchange Format 291
gradation 142, 252
gradation curve 174
graduated screens 150
grayscale resolution 251
grid cassettes 258
grid focusing 260
grid ratio 257
grid-controlled X-ray tube 115

H

half-value layer thickness 46
hardening equivalent 50
healthcare communication standards 288
heat storage capacity of the X-ray generator 203
heater plug filament 115
heating and cooling curves 203
heel effect 122
heterochromatic radiation 47
high voltage waveforms 52
high-frequency generators 106
high-kV exposure technique 41
high-kV technique 41
HIS, Hospital Information System 287
HL7, Health Level 7 288, 290
homogeneity 47

313

Index

honeycomb grid 256
HVL 46

I

I.I.-TV imaging chain 159
icons 215
ICRU radiation quality 47
IHE, Integrating the Healthcare Enterprise 292
image integration 240
image intensifier television technique 158
image intensity transformation 238
image quality 244
image sharpness 246
image viewing systems 179
imaging planes 264, 270
imaging plate 153
impedance scanning 75
indirect flat detector systems 133
indirect imaging 133
information science, medical 277
inherent filtration 48
inherent veil 139
intensifying screens 146
intensity 30
interlaced scanning 166
interregional networks 278
interspace material 257
interventional radiology 78
invert 236
iodine contrast 42-43
ion dose 58
ionization 28
ionization chamber 192
ionization current 54
ionization effect 33
IONTOMAT 199
iris collimator 124
isocenter 268
isotopes 23

J

JPEG, Joint Photographic Expert Group 291
jukebox 285

K

kerma 59

L

LAN, Local Area Network 278
landmark 236
LAO 265
laser imager 276
laser printer 276
last image hold 225
LCD monitor 180
lead equivalent 50
leakage radiation 122
light localizer 124
limiting resolution 255
line spectrum 30
liquid-bearing technology 111
liquid-metal liquid bearing 111
local dose 60
local networks 278
logarithm of the image signal 239
long-term archiving 285
look-up table (LUT) 239
low-kV exposure technique 41
luminescence effect 33

M

magazine technology 136
magnetic navigation 95
magnetic resonance imaging 74
magneto optical disks 224
magnification 236, 255
magnification, geometric 253
mammography screens 151
MAN, Metropolitan Area Network 278
mAs product 190
mass 22
mass attenuation coefficient 37
mass coverage 37
mass number 23
matrix 222
max./min. opac 241
measurement chamber 53
measuring stations 96
medical information science 277
mobile generators 108
mobile X-ray generators 83
modulation transfer function 250
monitor, CRT 180
monitor, LCD 180
monochromatic radiation 39, 47
motion detector 228
moving grid 258
multiformat camera 276
multileaf collimator 123
multi-line grids 258
multimap 242
multipulse generators 53, 106

N

nat/sub 235
native imaging 77

314

natural radiation 56
NEMA, National Electrical Manufacturers Association 288
networks, interregional 278
networks, local 278
networks, regional 278
networks, structures 279
neuroradiology 94
nuclear medicine diagnostics 76
nuclear reaction 34
nuclear spin 24
nucleons 23
nuclide 23

O

oblique projection 266
offline 225
online 225
operating types of DR 225
optical density 139, 157
optical density range 175
optical disks 224
optical focal spot 119
optical imaging processes 76
organ programs 194
oscillating grid 258
overlay fade 230
overtable fluoroscopy systems 88

P

PACS, Picture Archiving and Communication Systems 282
pair production 34
panoramic tomography 267
parallel grid 257
patient data management 275
patient record 276

performance data of X-ray tubes 200
Periscanning 231
Peristepping 231
Perivision 231
person responsible for radiation protection 71
personal dose 60
photographic effect 33
photon emissions 28
pixel 222
pixel shift 236
pocket dosimeter 69
point table 205
pre- and post-compare 231
prefiltration 49
primary collimator 123
primary quantum number 23
programmed radiographic technique 194
projections 264
protons 22
pulsed fluoroscopy 227

Q

quality factors 59
quantification methods 243
quantization noise 252
quantum noise 252

R

radiation dose monitoring 69
radiation exposure 55
radiation protection 64
radiation quality 46
radiation weighting factor 59
radiographic imaging 34
radiography systems 105
radiological positioning method 274

radiological safety officer 71
RAID, Redundant Array of Inexpensive Disks 285
RAM, random access memory 224
RAO 265
rare-earth screens 149
Rayleigh scatter 36
readers 155
reciprocity law failure 145
reference image 242
regional networks 278
remask 237
reporting workstation 284
resolution 152
resolving power 249
ring topology 280
ripple 51
RIS, Radiologic Information System 288
roadmapping 230
roam 236
roaming 241
roll film camera 169
rotating anodes 117
rotating-anode starter 120
rotation angiography 233
rotor 120
routers 286

S

sagittal beam path 271
sampling theorem 220
scanners 155
scanning process 166
scatter 35
scatter radiation 256
scintillator 172
selective dominant measurement (SDM) 197
selectivity 259
selenium detectors 175

315

Index

selenium technology 176
semiconductor effect 33
semiconductors 173
sensitivity classes 148
sensitivity index 141
sensitivity of film-screen system 148
sensitometers 141
sequential CT 101
servers 225
sheet film cameras 168
SI units 63
signal-to-noise ratio 168, 252
simultaneous multi-section method 269
skin dose 60
solarization portion 140
sonography 74
source-image distance 253
source-image distances, thresholds 261
space-charge effect 116
spatial filter technique 238
spatial resolution 249
spectrum 151
spiral CT 102
spiral liquid bearing 111
spiral-wound filament 115
spooler 286
spotfilm 125
stands 127
star topology 279
stationary anodes 117
stator 120
stem radiation 121
stereo technology 270
Straton 112
Straton X-ray tube 114

structure noise 252
summation 240
superposition 265
symbols 215
syngo 217
system components 104

T

teleradiology 287
television camera 159
terrestrial radiation 56
text functions 244
thermal focal spot 119
TIFF, Tagged Image File Format 291
toe 139
token ring 280
tomography 266
topologies 279
transparency 190
trauma and emergency X-ray systems 82
tube load computer 204
tube nomogram 201
tube voltage 51
TV iris diaphragm 159, 164
TV tube 165
twisted-pair cable 281

U

undertable fluoroscopy systems 87
uninterruptible power supplies (UPS) 184

V

valence electrons 23
vascular tracing 241
veil 142
video loop 226
vignetting 164

W

WAN, Wide Area Network 278
Wehnelt cylinder 114
Wehnelt electrode 116
winchester disk 224
windowing 237
workflow 282
workstations 182

X

xeroradiography 152
X-ray film 137
X-ray film cassette 134
X-ray film processing 145
X-ray generator 105
X-ray spectrum 30
X-ray systems 79
X-ray systems for bedside images 83
X-ray systems for computed tomography 99
X-ray systems for ESWL 92
X-ray systems for internal medicine 86
X-ray systems for mammography 84
X-ray systems for special angiography 93
X-ray systems for surgery 97
X-ray systems for the skeleton and chest 80
X-ray systems for urology 90
X-ray tube 108, 110
X-rays 28

Z

zoom 236

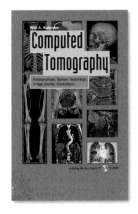

Willi A. Kalender
Computed Tomography
Fundamentals, System Technology, Image Quality, Applications

2nd revised and enlarged edition,
June 2005, approx. 250 pages,
approx. 80 illustrations, approx. 16 tables,
ISBN 3-89578-216-5
Approx. € 54.90 / sFr 81.00

The book offers a comprehensive and user-oriented description of the theoretical and technical system fundamentals of computed tomography (CT) for a wide readership, from conventional single-slice acquisitions to volume acquisition with multi-slice and cone-beam spiral CT. It covers in detail all characteristic parameters relevant for image quality and all performance features significant for clinical application. Readers will thus be informed how to use a CT system to an optimum depending on the different diagnostic requirements. This includes a detailed discussion about the dose required and about dose measurements as well as how to reduce dose in CT. All considerations pay special attention to spiral CT and to new developments towards advanced multi-slice and cone-beam CT.

For the 2nd edition many sections of this book have been updated. In particular, material on new x-ray technology, on 64-slice spiral and cone-beam CT scanning have been added.

The enclosed CD-ROM again offers attractive case studies, including many examples from the most recent 64-slice acquisitions, and interactive exercises for image viewing and manipulation.

This book is intended for all those who work daily, regularly or even only occasionally with CT: physicians, radiographers, engineers, technicians and physicists. A glossary describes all the important technical terms in alphabetical order.

Contents

System concepts · System components · Image reconstruction · Spiral CT · Multi-slice spiral CT · Dynamic CT · Quantitative CT · Image quality · Spatial resolution · Contrast · Pixel noise · Homogeneity · Routine and special applications · 3D displays · Post-processing · Quality assurance.

www.publicis-erlangen.de/books

Arnulf Oppelt (Editor)

Imaging Systems for Medical Diagnostics

Fundamentals, technical solutions and applications for systems applying ionization radiation, nuclear magnetic resonance and ultrasound

2nd revised and enlarged edition,
September 2005, approx. 760 pages,
ISBN 3-89578-226-2
Approx. € 119.00 / sFr 176.00

The book provides a comprehensive compilation of fundamentals, technical solutions and applications for medical imaging systems. It is intended as a handbook for students in biomedical engineering, for medical physicists, and for engineers working on medical technologies, as well as for lecturers at universities and engineering schools. For qualified personnel at hospitals, and physicians working with these instruments it serves as a basic source of information. This also applies for service engineers and marketing specialists.

The book starts with the representation of the physical basics of image processing, implying some knowledge of Fourier transforms. After that, experienced authors describe technical solutions and applications for imaging systems in medical diagnostics. The applications comprise the fields of X-ray diagnostics, computed tomography, nuclear medical diagnostics, magnetic resonance imaging, sonography, molecular imaging and hybrid systems. Considering the increasing importance of software based solutions, emphasis is also laid on the imaging software platform and hospital information systems.

Contents

Principles of Image Processsing: Physiology of Vision · Subjective assessment of image quality · Image Rendering · Image Fusion · Navigation.

Physics of Imaging: X-Ray and Gamma-Radiation · Magnetic Resonance · Ultrasound · C-Image Reconstruction · System Theory · Reconstruction Algorithms.

Imaging Instrumentation: Displays · X-Ray Diagnostics · Computed Tomography · Nuclear Medicine · Magnetic Resonance Imaging · Sonography · Special and Hybrid systems · Molecular imaging.

Information processing and distribution: Imaging Platform (Syngo) · Computer Aided Diagnostics · Hospital Information System.

www.publicis-erlangen.de/books